REJOICE IN ADVERSITY, TRIUMPH IN WAR

Rejoice in Adversity, Triumph in War by Major General Rajpal Punia, YSM is a compilation of interesting military anecdotes. This book covers the most eventful military tenure of Major General Rajpal Punia.

I hope the book will not only acquaint the young generations with the valour and bravery of the Indian Army, but will also inspire them, kindling in them the passion to serve their motherland. I congratulate him for this wonderful compilation and wish him success.

**Hemant Soren, Chief Minister of Jharkhand,
Ranchi, Jharkhand**

The value of memoirs written by men who have done their service in the ranks has been well received and recognized. Their value well merits the attention of all interested in the glorious history of the Indian armed forces. The numbers

Celebrating
30 Years of Publishing
in India

and quality of such reminiscences clearly serve as part of the knowledge base for future.

The Indian armed forces, be it the Army, Air Force or Navy, all have their unique identity and have brilliant records in safeguarding the country's unity, integrity and sovereignty. United, they built a reputation for which every Indian feels proud of them. One looks up to them not only to protect our nation, but also for their impeccable qualities that make them stand out from everyone else.

Gallantry has always commanded respect and recognition. The country has successfully fought several wars, leaving behind a saga of unmatched valour and gallantry of our armed forces. *Rejoice in adversity, Triumph in War*, apart from providing an exhaustive overview of the military culture, aims at providing civilians with a basic understanding of the unique life and culture that is the military. The author is a veteran of a number of military campaigns/operations during a very eventful military tenurem and he is extremely proud of his military record. With a diverse and extensive range of fighting and commanding experience, he understandably has some extraordinary stories to share. The military anecdotes that he has mixed with life lessons would make reading interesting.

I wish the book to be enriching and enlightening.

Ganeshi Lal, Hon'ble Governor of Odisha
Bhubaneswar, Odisha
20 December 2021

I am thrilled to have the opportunity to praise a book by a brave soldier, who wrote epic tales in Olive Green, both inside and outside the country. Major General Rajpal Punia played a significant role in spreading the reputation of the

Indian Army internationally. This Book covers some of the most memorable moments of General Punia's military and personal life.

General Punia's *Operation Khukri* has excited readers and made them proud of our Army. It tells the story of the challenges and survival of a small group of Indian soldiers who were deployed in Sierra Leone, Africa, as part of the United Nations peacekeeping assignment. General Punia led the Indian Contingent of 233 soldiers, who were taken hostage by the rebel Revolutionary United Front in Sierra Leone. Surrounded by the enemy in the remote village of Kailahun, his adventurous and timely actions saved the lives of all the soldiers in the barracks.

As a reader, I think *Rejoice in Adversity, Triumph in War*, like *Operation Khukri*, will provide readers with a heartwarming reading experience. The language of his book is simple and heartfelt; it will easily captivate the reader. It begins with General Punia's experiences during his Sainik School days. He comes from a completely non-military background. His father was a dedicated farmer. He wanted to see his two sons as military officers. Unfortunately, only one of the children got selected due to technical reasons related to age. According to interviewers, General Punia was chosen by his father. When he later asked his father about this, the latter smiled and said, 'I had seen the General in you in your childhood.'

For more than a decade, Military school and college life transformed General Punia from an ordinary peasant's son to a courageous soldier. Teachers, trainers, seniors, friends and different living environments shaped General Punia as a soldier and as an individual. But what turned him into an invincible army officer seems to be his attitude of refusing to surrender in the face of a crisis.

The author recalls a handful of eventful military experiences that he and his colleagues had on the Indian soil and abroad, which includes being part of Kargil Operation Fighting Breakout to extricate 233 peacekeepers deployed in a UN peacekeeping assignment as part of Operation Khukri, commanding 450 km of Line of Actual Control (LAC) in Arunachal Pradesh, encompassing a number of disputed areas, with frequent patrol clashes with the Chinese, and finally, while commanding the prestigious Armoured Division vacating Dera Sacha Sauda at Sirsa. This memoir is not just a description of war experiences, but it is also a combination of sweet, bitter and nostalgic laughter. As a child, he was fascinated by the shot put competition and its eight-pounded iron ball, his years of constant defeat to his senior, Dharamvir, in the shot-put competition, and the ridicule of Dharamvir and his friends. These will touch the readers' hearts. General Punia was fight more who refused to surrender from his childhood. Defeats only made him a fighter and work harder. At the last opportunity that he got, General Punia defeated his undefeated opponent! Many things come to mind, such as the personal attachment with loving 'Baba' at the Military Academy, beautiful days in Wellington, the laughing episode at Boxing training conducted by his colleague at Hotel Riga … I do not extend my words for the fear of breaking the thread of the readers' curiosity,

I am presenting this book to the readers with immense pleasure, wishing General Punia to come back to readers with more books like this.

Jai Hind!

P.S. Sreedharan Pillai, Hon'ble Governor of Goa
Dona Paula, Goa
3 January 2022

REJOICE IN ADVERSITY, TRIUMPH IN WAR

A MILITARY MEMOIR

RAJPAL PUNIA

HarperCollins *Publishers* India

First published in India by HarperCollins *Publishers* 2023
4th Floor, Tower A, Building No. 10, DLF Cyber City,
DLF Phase II, Gurugram, Haryana – 122002
www.harpercollins.co.in

2 4 6 8 10 9 7 5 3 1

Copyright © Major General Rajpal Punia 2023

P-ISBN: 978-93-5699-018-0
E-ISBN: 978-93-5699-019-7

The views and opinions expressed in this book are the author's own and
the facts are as reported by her, and the publishers are not in any way
liable for the same.

Major General Rajpal Punia asserts the moral right
to be identified as the author of this work.

All rights reserved. No part of this publication may be reproduced,
stored in a retrieval system, or transmitted, in any form or by any means,
electronic, mechanical, photocopying, recording or otherwise,
without the prior permission of the publishers.

Typeset in 11.5/15 Adobe Caslon Pro at
Manipal Technologies Limited, Manipal

Printed and bound at
Thomson Press (India) Ltd

This book is produced from independently certified FSC® paper
to ensure responsible forest management.

To my father …
Whatever I am today, you had sacrificed for yesterday!
I hope you're proud wherever you are. This is my first book after you left us. Thank you for being that force that helped me realize my dream. It is only because of your sustained efforts that I could join Sainik School.

Also, my children Abhishek, Arjun, Damini and Noddy.
Thank you for bringing so much joy in my life!

CONTENTS

1   A Dream Too Small   1
2   Charity Must Begin at Home   17
3   Jaadu Ki Jhappi   27
4   Adapt and Stop Smothering Captain Smokey   41
5   Rajdoot Outside the Box   53
6   Hurt Your Ego to Progress in Life   63
7   The Curious Case of the Foot   71
8   Flag Hoisting on a Sand Dune   81
9   The Bus that Changed My Life   91
10   Good Old Soldier Simar Singh   101
11   Breaking the Thread with his Chest   109
12   Silver Jubilee of a Grenade Splinter   123

Contents

13	Boxing at the Riga	131
14	Bonhomie with the Intelligence Warrior	139
15	Bonjour at the Guinean Border	149
16	Soldiering at the Cost of Friendship	159
17	Sanghe Shakti on Balwinder's Farm	167
18	Sojourn in the Mecca of Management	177
19	Goba The Hunter in the Fantastic Fifth	187
20	Scout Camp at Fifty on an Island	197
21	The Soldier Who Vacated Dera Sacha Sauda	207
22	Hasten Slowly	223
	Acknowledgements	231

1

A DREAM TOO SMALL

My life in uniform started quite early, at the tender age of ten in the year 1973, when I was admitted into a Sainik School in Chittorgarh, Rajasthan. It was a boarding school with all the trappings of a military environment, which played a significant part in moulding my future life and career in the Indian Army.

The Sainik Schools were introduced in the year 1962. The idea was conceived by the then Defence Minister, V.K. Krishna Menon, to rectify the growing regional and social imbalance among the officer cadres of the Indian Armed Forces. Since the colonial days, the officer cadres of the Armed Forces were generally populated with the families of the royalty and regional satraps who had access to quality public school education and exposure as part of the higher strata of the social echelons. Also, most of the officer cadres were generally drawn from the few regional and accepted martial societies, though the men they commanded represented almost the entire Indian subcontinent. Consequently, as the nation continued to grow as a young democracy, a perceptible disconnect was being observed between the officer cadres and the men they commanded, a fallout of which was seen in the 1962 Indo-China War. To achieve a broader recruitment base, and also bring in a more representative hue among the officer cadres, the Sainik Schools were established with the objective of having at least one school in each state to act as a feeder organization to the National Defence Academy at Khadakwasla and the Indian Military Academy at Dehradun. The schools were modelled in line with the best comparable public schools of those times in our country, and the educational costs were hugely subsidized by the government to make it affordable for even the weakest economic sections of the society who aspired to send their wards to these schools. The entry criteria were strictly based on merit, and it was indeed an arduous task to acquire a seat in a Sainik School.

I come from a completely non-military background; my father was a dedicated farmer. But it had always been his dream to see both his sons, that is, my elder brother and me, as commissioned officers in the Indian Army. He had planned our initial academic pursuits accordingly, and after a short stint in a training institute, both my brother and I qualified the written examination for Sainik School, Chittorgarh. The news was greeted with jubilation by the entire family. Clearing the written examination was only the first hurdle, and we still had to navigate our way through another one in order to be a part of the prestigious Sainik School. The selection process consisted of the written examination, followed by a personal interview with the principal of the school, who was a serving officer of the Indian Army in the rank of a Colonel. As the days left for the interview reduced by half, the acceleration of our efforts doubled. Our discipline was checked every morning by our father, who ensured that not even a strand of hair swayed away from its place. My father ensured that both my brother and I conversed only in English so that we would have enough words to use during our interview.

Life in those days was in stark contrast to the present. Forget malls, there wasn't any concept of readymade garments. The local tailor was the designer, and indistinguishable bespoke garments were stitched from bolts of fabric for all children. Hence, matching bush shirts and shorts were stitched for my brother and me, and donning crisp identical outfits, accompanied by our father, we boarded the first train at the break of dawn for Chittorgarh. There weren't any direct trains during those times, and hence, we changed the narrow-gauge train at Jaipur. The house of forts and fortresses, Chittorgarh was a splendid experience for us. Our cycle rickshaw was traversing through the narrow lanes that were reverberating with the legendary tales of the heroism of Rana Sangha and the obstinacy of Allaudin Khilji. The great walls of the Chittor fort, protectively overlooking the city, exuberated with pride and reverence. After our little voyage, we finally halted in front of a colossal bi-parting swing iron gate that was the doorway to my scholastic sanctum, the terminal station of my toil train. Through the apertures of the iron poles, I could faintly discern our Tricolour tied to a tall white mast fluttering against the ineffably blue sky. The towering, white-walled school building, in the shadow of the trichromatic national flag, looked regal and majestic. As I walked through the high-ceilinged semi-oval corridor, my feet started trembling. I was walking into my dream location; this was the starting block of the Jenga of my dreams and I had to conquer it. I couldn't let it dwindle! I noticed pictures of the school's august alumni adorning the side walls. It gave me an immense sense of pride that I might get an opportunity to sit on the same benches that were once occupied by such distinguished legends. We were escorted into the waiting room filled with a bunch of other candidates. There was angst in the air as we occupied the scarlet straight-backed chairs stationed on the side. I went into the interview room right after my brother. The principal could relate me to my brother as our outfits bore testament to our tribe. While different general awareness issues were whirling in my mind, the principal chose to target my date of birth at the outset. Now, you all must be wondering as to what could possibly be wrong with someone's date of birth! It was problematic as the age difference between my brother and me was merely six months on paper, something that was preposterous on another level. Six months of age difference between siblings is a biological impossibility, and hence, we were caught. In those days, there weren't any birth certificates, especially in the rural parts of the country, and ordinarily the school teacher would document the age of the child. In our particular case, it was a faux pas committed by our class teacher. While my elder brother was older by more than a year to me, the class teacher had marked his date of birth in such a way that officially we were just six months apart. Anyway, my father was summoned to the interview hall and was given the option of admitting one child into the Sainik School, as it was evident that the other one was either overage or underage for admission. My father didn't even think for a second before putting his hand on my head, indicating his choice to the principal. Thus started my tryst with destiny in a life filled with such moments where small and apparently insignificant actions of mine or somebody else have gone on to make a huge difference in my later years. I had later on asked my father the reason why he had selected me for admission in the Sainik School over my brother. My father had smiled and said, 'I had seen the General in you in your childhood.'

And that is how destiny is made at times. Today, when I reminiscence about that day, I always make it a point to ask my father what prompted him to put his hand on my head so quickly, and my father always smiles and exclaims that he *knew* the 'General' in the making. Thanks to my father's judgement, I joined the Sainik School at the age of ten, way back in the year 1973.

After completing all the necessary formalities, I was admitted into Sainik School, Chittorgarh. I still vividly remember the day my dream became a reality! I was so engrossed in dancing in my dream bubble that I became completely oblivious to reality. My admission meant that my father would leave. I walked up to the main gate alongside my father. Throughout those 200 metres, I kept trying to hold my father's hand, but I just couldn't. I was scared but didn't want to admit it. My father was emotional but didn't want to express it. As we reached the gate, my father turned towards me one last time and said, 'Beta, maine apna farz poora kar diya hai, ab yeh tumhara waqt hai kuch karne ka' ['Son, I have fulfilled my duty, now it is your time to shine']. An otherwise extremely unemotional man stood before me in a sombre state. He wanted to conceal his tears, and hence, left in an abnormal hurry and didn't even turn back one last time. There I was, a ten-year-old, standing across the gate, alone, scared, crying. I wish I could have hugged him. I couldn't walk back towards my room. It was as if my feet were tied with the emotional weight of the separation from my father. With trembling legs and watery eyes, I somehow managed to reach my room. That night felt like eternity as I kept rolling left and right. I pulled out my parents' photograph that I had tucked under my pillow with love. The rectangular photo paper was my comfort blanket as I tried to sleep.

Life in the Sainik School is an endless regime of physical training, academics, games, sports, and a host of cocurricular activities, in which all the cadets[1] are expected to participate. The following day, I woke up and something was amiss. My mother's sweet voice didn't wake me up; instead, the school bell jolted me back to my senses. The day started very early with a morning session of physical training, followed by the academic classes and organized games in the evening. Even in the later part of the evening, the cadets had to attend the preparatory classes under the supervision of teachers until the call for 'lights out' at night. I am glad that the routine was so chockablock because after the initial settling in, I was so engrossed with it that emotions took a backseat. In the midst of all things academic, there were all kinds of competitions and tournaments which were played. All the cadets competed with immense zeal and enthusiasm as the pride of their respective houses were at stake. The school had the best of infrastructure and facilities, comparable to the best public schools in the country. We were served the most nutritious meals in the Cadet Mess, and even our daily requirements

---

1   Students of Sainik Schools are referred to as cadets.

like laundry, barbershop, healthcare, etc., were taken care of. For someone like me, who came from a completely rural background, the facilities offered by the school were a major novelty.

As a novice, I was put in the Holding House.[2] During the initial days, the Holding House used to be quite a forlorn place, with most of the cadets moving around with wet eyes and sniffing noses, missing their parents and homes. We banked on each other, searching for mutual comfort, and it was in these rooms and halls that some of the most lasting and lifelong friendships were forged. One such friend of mine was Govind Singh Rathore, the son of a serving soldier. His father was posted in the prestigious Indian Military Academy at Dehradun. He had become quite a buddy and we used to have our gossip sessions lying on our beds, well after 'lights out' was called. During one such whispering campaign, Govind casually asked me, 'Rajpal, what is your dream in life?'

His question caught me off guard. I had no answer because I had never really given it a thought. So instead, I asked him, 'Govind, you tell me, do you have a dream?' He was quick in answering, 'My dream is to win the Sword of Honour[3] during the passing out ceremony in the Indian

---

2 Cadet hostels are called Houses in Sainik Schools.
3 The Sword of Honour is an award and honour bestowed during the final passing out ceremony at the Indian Military Academy to the cadet adjudged best in merit in the whole course.

Military Academy.' I was really impressed with Govind's dream, and thereafter, it became a priority for me to formulate my own personal dream.

Within a few months of our joining, the annual sports meet of the school was organized, in which the cadets participated and competed in various track and field events. As a wide-eyed youngster, I was mesmerized seeing the various events taking place and the way our senior cadets participated in them. In those days, the senior cadets used to be like heroes and idols for us, and we used to be fascinated by their achievements and on-field mannerisms, which in fact many of us would try and emulate. One of the most absorbing events was the shot put, where a player was required to throw an iron ball to the farthest possible point in order to win. One cadet, Dharamveer Suhag, was winning the event hands down by throws well exceeding his closest rivals by metres. Even though he was still in the junior section, Dharamveer was an excellent sportsperson and had the built and body to match his sporting prowess. He was the grandson of the horse-riding instructor of our school. Even his elder brother, Dalbir Suhag, was a senior cadet of the school who went on to become the Chief of the Army Staff of the Indian Army. Dharamveer's mannerisms, the way he pivoted to throw the shot put, along with his vocal exultations after the throw, were almost at a mythical level; the entire school used to cheer him on during his performance. He was indeed the star performer on the sports field. Thus, watching him perform, my dream was born. That night during my ritual late-night gossip with Govind, I shared my dream with him. I said, 'Govind, I, too, have a dream now, and my dream is to beat Dharamveer in the shot put event.' Govind was initially silent and then broke into a raucous laugh. He called me crazy and asked me to go to sleep. However, I could not sleep at all that night. My fledgling dream kept nagging on my mind as I kept devising ways to go about achieving my dream.

Early next day, before anybody had woken up, I took a bath and went to the school temple that was located near our hostel and made the sincerest prayer a ten-year-old could. I prayed to the Almighty to help me succeed in my newly created dream; a difficult feat to achieve indeed. From that day onwards, I would try to run faster during our morning physical training, use our water bottles as weights and practice after everyone else had gone to sleep. Soon, the school went on vacation, and we all went home for our summer holidays. Having been away from home for the first time, my homecoming was nothing less than a festival, and my mother took it upon herself to pamper me with all the goodies possible. One night, I confided in her my dream. Though she had no idea what shot put was, she simply took it as something that had to be done to fulfil my dream and she immediately instructed my father to procure a shot put at the earliest. My father, who was well versed in sports and sporting matters, initially had a hearty laugh at the thought of his ten-year-old son wanting to start practising shot put. He even consulted a local athletics coach in our neighbourhood who, too, advised against my trying out shot put, a sport requiring major muscular developments, so early in life. However, my mother put her foot down and said that if her child wanted to purchase and practise the shot put, he would get it.

So, with some money from my mother, I, along with my friends, purchased a shot put weighing eight pounds. Thereafter, I started the routine practice along with my friends in the local ground. Sensing my dedication, the coach came forward with his support by way of guiding me with the correct throwing technique. Whatever the coach was teaching me in the evening, I was religiously practising the next morning. The iron mass of eight pounds was my most valuable possession and I would sleep with it every night. My mother increased my milk intake, and later my father tied up with a teacher from our area to arrange for one kilogram of buffalo milk every day for me in the hostel. Our relatives who would see me throwing a massive iron ball would taunt my mother that I had gone mad. On all such occasions, my mother would ask them to mind their own business. Finally, at the end of my vacation, I went back to school, accompanied by my buddy and most prized possession, my eight-pounder shot put ball. The daily routine of the school did not permit me the opportunity to practise as much as I wanted to, and so I would wake up in the middle of the night to master the art in the hostel courtyard. On one such moonlit night, while I was busy practising, the Housemaster noticed me, and as luck would have it, he had a sporting bent of mind and understood my passion. Since athletics contributed to major points towards inter-house competitions, I was now allowed to practise in the morning as well as in the evening, irrespective of the other school commitments.

As the years passed, the efforts I was putting in started to show in the form of visible improvements in my physique. I started looking like an athlete. Yet, every year that I would compete in the annual shot put, I could only manage to secure the second position behind Dharamveer. I remember that once, out of sheer frustration, I had blurted out in front of Dharamveer that I would beat him one day, but it made matters worse because thereafter, he started ridiculing me in public in front of all my classmates. With every instance of humiliation, while my self-esteem was getting battered, my resolve grew stronger. Every vacation, my iron buddy was my best companion, and with time, the coach gradually started paying more attention to train me. He started guiding me to improve both my technique and my power behind the throw. Appreciating my seriousness, my parents and relatives started encouraging me. The funny thing was that by now everyone in my family had heard of Dharamveer, even though they had never met him. From my first year of participation till my last at the school, my performance in shot put had improved a lot and I kept inching towards Dharamveer. Yet, the coveted gold kept eluding me, and I continued to secure the silver medal on the podium while Dharamveer's ever-increasing taunts and jibes persisted.

School ended with both Govind and I qualifying for and joining the National Defence Academy at Khadakwasla. Dharamveer, who happened to be our senior in the school, also qualified in his third chance, and as a result, became our coursemate in the Academy. At the Academy, during the next three years, my chase for the shot put 'gold' continued, with no change in the overall position secured by me. Dharamveer was still the best, closely followed by me. The only difference was that the gap, which used to be in metres, had now reduced to inches. Dharamveer, along with Yogender Chahar, who was the bronze medallist in shot put, would not spare any opportunity to taunt and belittle me in front of everyone. Dharamveer, in particular, would boast and narrate in front of all our coursemates that how, despite my best efforts for the past so many years, he still was unbeatable. This continued throughout my three years at the National Defence Academy. After our graduation, we moved to the Indian Military Academy, Dehradun, for the final one year of our training. I realized that this was my last chance to beat Dharamveer, and thus, I further invested my time and energy in the practice. I soon found that due to the jam-packed training schedule, the only time available for practice was at night. So, I was busy at night practising my shot-put throws. While I kept on dreaming and working towards the elusive gold in shot put, Govind was working steadfastly towards his dream and was already being perceived as the most likely recipient of the coveted 'Sword of Honour'.

The designated day for athletics championship at the Academy finally arrived. The night before the championship, I could not sleep at all and kept on tossing and turning in my bed. The next morning at the break of dawn, I was the first person to reach the ground. The entire stadium was empty as the sunbeam had not yet broken the dark of the night. Taking advantage of no one being around, I went up to the cemented circle from where the shot put is thrown and quietly prayed. Soon, the stadium started filling up; and since my rivalry with Dharamveer was by then known to one and all, there was a tremendous amount of curiosity and interest among the Gentlemen Cadets.[4] The event started with all of us giving our best shot, and suddenly I realized that I was on my third and final attempt. Everyone had completed their throws and my best till now was still a few inches short of Dharamveer's best. I now had a sudden sensation of hundreds of eyes fixed on me. I closed my eyes, and in my mind, I could still hear the grunting sound of Dharamveer along with glimpses of my successive defeats, the taunting, and the ridiculing. It is difficult to describe what happened next, as I barely remember it. I was suddenly jolted back to my senses with a loud cheer in the stadium. I opened

---

[4] The trainees at the Indian Military Academy are referred to as Gentlemen Cadets.

my eyes and realized that I had already thrown the shot put and it had cleared Dharamveer's best by quite a distance! While the cheering was still on and a few of my coursemates were carrying me on their shoulders, I was numb with ecstasy as well as fatigue and I could not hold back the tears rolling down my eyes. Today, after almost three-and-a-half decades, while I continue to cherish and be nostalgic in the memory of that achievement; when I look back, I cannot help but smile at my 'DREAM'.

But today, as I reflect on that part of my life, I start questioning myself: Was the whole effort worth the dream or was my dream too small to be worth the effort? What if I had the opportunity to go back in time and confront my own self that night as I was about to confess to my dear friend Govind that 'I had a dream.' I would have probably advised myself to have a 'DREAM WORTH DREAMING'. While it is extremely important for all youngsters to start dreaming, it is equally essential that the dream should be the biggest possible dream so that the best possible years spent towards achieving the same do not end up looking a tad futile at the end. I invested almost ten years of my life, my time, my focus, and my energy in beating Dharamveer in shot put. Though I ended up achieving this dream, I wish from the core of my heart that someone should have guided me towards a bigger dream in life. Even if shot put was my dream, it shouldn't have been limited to just beating Dharamveer; rather, it should have been to be the best in the business in the entire world.

Probably an Olympic gold in shot-put would have been the ideal dream.

Yet, I would not disparage my experience of dreaming completely. In the end, I thank Dharamveer, who went on to be the best of my friends, for being my role model and my motivation towards achieving excellence, no matter how inconsequential the achievement may feel today. Even today, on any difficult day, as I look at the vintage photograph of the prize distribution ceremony where I stood on the podium at the number one position and Dharamveer stood at number two, even without realizing, a faint smile passes my lips and my eyes shine bright, exactly as it shone when I had opened my eyes and heard the entire stadium cheering for me!

2

CHARITY MUST BEGIN AT HOME

Dr Stephen Covey, in his book, *Seven Habits of Highly Effective People*,[5] has spoken at length about the need to develop the capability to manage oneself within the given environment and how this skill or capability can only be developed through an organic growth from within one's immediate family and milieu. Hence, all our social, emotional and spiritual

---

5   Stephen Covey, Seven Habits of Highly Effective People (Mumbai: Free Press, 1989).

management ultimately get shaped through our life's experiences, guided and moulded through our myriad experiential anecdotes and situational reactions.

My life as a soldier and a human being can probably be summed up as a continuum of managing my surroundings and reacting to the given situations in such a way that the outcomes and their effects on me and my socio-professional neighbourhood can be termed more or less positive always. My initial years of training right from my days in the Sainik School up to three years in the National Defence Academy and the year-long training at the Indian Military Academy had a huge role to play in this. And not just my teachers, formal instructors or trainers, but to a large extent, my seniors, my peers, and my circumstances, through numerous challenges and situations, developed in me the ability to manage my life skills and utilize them to my advantage. The story of my life is indeed a tale of how to deal with and manage every adversarial curve ball thrown at me by fate. During my school days, travelling to and from Sainik School in Chittorgarh to my hometown was a herculean task and an adventure in itself, requiring the most innovative managerial and survival skills. To reach school, one had to take a bus to Jaipur, and then from there one boarded the famed Chetak Express train to Chittorgarh. The trains in those days were hauled by steam locomotives and had very few reserved compartments. Whatever few reserved compartments were there were generally always booked right from the originating station of the train in New Delhi, and I don't remember a single instance when, in spite of my father's huge efforts, I could ever manage to travel on a reserved ticket or berth during my entire time in the school. Consequently, the only way out was to travel in the unreserved class. I remember even now how, a couple of days prior to the culmination of our term break, the rumbling sound of the train rolling down the tracks would wake me up instantly and the 'Chetak Fever' would haunt me like a bad dream. The train would reach Jaipur in the middle of the night, and after trying in vain to enter every compartment, I would have no choice but to climb the roof of the bogie. On such occasions, one felt like Maharana Pratap[6] mounting his Chetak![7] Our school was located at the same ground where Alauddin Khilji's forces had camped in the year 1303 AD, while he, obsessed with Queen Padmavati, was trying to persuade the Rajput ruler Maharaja Ratan Singh to allow Khilji a glimpse of the beautiful queen. Later, both the forces fought on the same ground, where centuries later our school came up. It was a beautiful location on the banks of River Berach, right in front of the massive and most formidable Chittorgarh fort. It used to be a fascinating sight to witness the sun rising from behind the fort and the sun rays reflecting from the river on to our school like the blessings of God Almighty Himself on all the

---

6   A legendary king.

7   Maharana Pratap's horse.

students. Despite the school administration putting the area around the river out of bound for all the students, the very first lesson of swimming was imparted to us by our seniors in River Berach. Bathing in the river used to be more of a compulsion than an option, since the water taps in our hostel bathrooms were only busy making musical sounds, occasionally gracing us with some sprinkles of water, and one had to be really lucky to match his timings with those sprinkles. The watermelons on the banks of Berach used to take care of our appetite, and they were so sweet that even today, whenever my wife brings home the fruit, I remember my favourite Berach! There were many sugarcane fields next to the river, and the owners always ensured a proper guard since the reputation of the school students had spread to the adjoining villages like wildfire, thanks to the many exploits and feats of our seniors in the past. We had to come up with an out-of-the-box solution for our 'silly mischief', since the canes were too juicy and tempting to resist, and we convinced ourselves that as future military leaders this was only a test in the planning and execution of a perfect plan. Thus, a brainstorming session was conducted, and a plan finalized; a night operation was considered to be the best option; the timing of the operation had to be synchronized with the goods train rolling over the bridge on River Beach. This would ensure that the sound made by us cutting down the sugarcane would be suppressed under the sound of the goods train passing over the iron girder bridge.

I am sure you now understand how much we owe to our alma mater as military professionals! Our seniors would humbly take on the responsibility of training the future responsible citizens of our country. Besides many other such training sessions, the most spine-chilling leadership training sessions would invariably be conducted on New Year's Eve. There were a couple of empty barrels in our hostel courtyard, which used to be filled with ice- chilled water, and all of us were forced to take a dip in the barrel every ten minutes while facing the Chittorgarh fort. This used to continue till the clock struck twelve on 31 December. On every such occasion, we would desperately miss the warmth of our mother's laps since we were still children, after all, yet to step into our teens. These were testing times, the only silver lining being the fact that we had suddenly matured far beyond our years. One can say, however, that this maturity came at a greater cost; we lost our childhood! We had aged before time. Children our age were enjoying the comforts of their homes while we were roughing it out, facing the wrath and fury of our seniors, who seemed to really believe in that old saying, 'Spare the rod and spoil the child.' We were more like a means for our seniors to pass their time. Despite the challenges and hardships of hostel life, there were many takeaways. Survival and management became our mantras. I was lucky to have made the grade in my very first attempt for the

National Defence Academy,[8] which meant that I was barely sixteen when I joined. Still in my teens, challenges continued to confront me in their varied forms as part of the academy routine. Time management in its ultimate form was learnt during a thirty-minute break after some rigorous and back-breaking physical training and drill classes early in the morning. The things I had to do during these precious thirty minutes included cycling around two kilometres to reach my room, having a quick bath, having my breakfast in the Cadet Mess, and finally, cycling again a couple of kilometres to reach the academic block for my next class. All the activities mentioned above were sequential, and it was very difficult for one to avoid any of these activities in between, except perhaps the breakfast, which was indeed impossible to avoid, since after the rollicking activities of the morning, the craving for a bite used to be at its peak. These thirty minutes used to be under the critical eyes of our senior cadets; they would be hounding us in case we violated our table manners during breakfast.

The first and most important lesson imparted to us by our seniors the moment we arrived in the Academy was 'to manage'. For every crisis or adverse situation, our seniors would only ask us to manage it on our own. Even if we lost our bicycle (trust me, that was a very common occurrence!), the seniors would simply say, 'Go

---

8   National Defence Academy, Khadakwasla, Pune.

and manage'. This chant was repeated by every senior in every crisis situation, so much so that having to learn to manage and survive in every situation that life would present became a part and parcel of our lives. I recollect one such critical situation, when one of our coursemates fell down from a horse during his riding class, and the horse vanished into a bush. All of us panicked, wondering who would retrieve the horse, and how the situation would be handled; and yet, the senior cadet present with us only asked our coursemate to manage the situation and asked the rest of us to proceed for the next class. The poor fellow, instead of dressing his wounds, was now running behind the horse; how he managed to retrieve it is anyone's guess! In retrospect, I can say that my days at the boarding school, and subsequently, at the Academy, while definitely not easy, were blessings in disguise. The basic instinct of survival was so deeply engraved in my young mind that it truly helped me in handling crises many years later, such as in the peacekeeping assignment in Africa, when we were surrounded by rebels for almost three months without any food or rations, which I have mentioned later in the book. I now realize that it was not the hostel life or the Academy days which taught me to handle crisis, but it actually was the independence of handling my own routine problems (which I had in plenty). Here, I would make a request to parents who, by being overprotective, are damaging their children. Be like the seniors we had in the Academy and simply ask your children to handle their own routine problems. Save your umbrellas for the rainy days, and let your children play in the sun (maybe, at times, even in the rain). You will see them grow and mature, and I can assure you, you would be doing yeoman's service by letting them stand on their own two feet!

Many years later, when I attended the prestigious Higher Defence Management Course at the College of Defence Management in Secunderabad, Telangana, the first lesson imparted to us on the very first day was to 'learn to manage yourself'. The same is also reinforced in Dr Stephen Covey's book. Unfortunately, individuals today overreact at the drop of a hat, and as a result, families are breaking up from joint to nuclear to single. The meaning of the word 'patience' has been completely lost. The irony is that when life was simple, elders were around as we all lived in joint families, but when life has become complicated today, elders are missing as it has become the era of nuclear families. Therefore, it is all the more reason to handle your emotions; 'self-management' assumes the highest priority. Let me finish with a few words of wisdom for today's parents. You must instruct your children with the 'Whats' of life rather than the 'Hows'. For almost every situation, there are always numerous ways to approach it, so let your children apply themselves fully. This will ensure that the child starts taking decisions early in life, and you as parents empower them to face the world. Do not act like a big tree under which a small tree can never grow. Make your child independent so that they can handle their own crises and be only what they really need—a guiding figure. The child would rather fall on their own than hold your finger while walking and never be able to walk by himself. Be like a mother bird, who teaches her little one to fly, unlike humans who, unknowingly, are the biggest impediments to the growth of their children, who keep carrying them on their shoulders instead of allowing them to fly. Remember to accept the child's mistakes because it is absolutely normal to make mistakes. Do not compare your child with yourself or with other children; let the child have their own identity. Give the child the love and affection they really deserve and let the little one grow in the warmth of your home and be your best buddy, companion, partner, and a friend! Remember that your child is already facing some of the most challenging times, when even one wrong step can ruin the child's life; so, who else can be a better guide than you? The only way a child can open up to you is if you ensure that the child is relaxed at home. If the child is not relaxed at home, then I question the entire institution of 'home'. The least you, as parents, can do is to make your child feel at peace, and that indeed is only possible once you yourself learn to be at peace! Therefore, indeed, 'Charity must begin at home'!

3

JAADU KI JHAPPI

Seven years at Sainik School, three years at the National Defence Academy, Khadakwasla, and finally, a culmination year at the Indian Military Academy at Dehradun were more than sufficient to transform a little brat into a disciplined and competent officer of the prestigious Indian Army. The Indian Army owes its origin to the British Army wherein ranks up to a non-commissioned officer are worn on the sleeves of one's arm, symbolizing hard work, while the officers wear their ranks on their shoulders, signifying the epitome of responsibility and accountability. The word 'accountability' conveys the real spirit and ethos of the Indian Army, in general, and of a gentleman officer, in particular. I do not think any other organization in the world trains individuals by putting them through eleven long years of strenuous scrutiny, sweat, and tough grind.

Now that our ceaseless training was finally approaching its finale, there was an air of excitement all around, with each one of us eagerly awaiting the shining golden star on our shoulders. The significance of the passing out parade, followed by the pipping ceremony, can be ascertained from the fact that the Academy invites the parents of all 300 gentlemen cadets to give them an opportunity to personally honour their children by pipping the star on their shoulders. In addition to the Academy's official invite, I had also sent a telegram home to my parents, inviting them to the Academy for my passing out parade and pipping ceremony. Our schedule during the last few days in the Academy kept us well prepared, with multiple practice days assigned for the passing out parade. Every foot stamp, every hand gesture and every head tilt had to be in absolute sync, which required a tremendous effort from the Gentlemen Cadets as well as the 'Ustaad',[9] who was permanently marinating in anger at the discordant cadets. It was all the more challenging for me, as I had been appointed as the 'Battalion Under Officer' responsible for the command of the 'Thimayya Battalion'; one of the four

---

9   Instructor.

battalions of the Academy during the parade. While the rest of the Gentlemen Cadets were to march together in unison carrying their respective rifles, the command of a battalion involved leading the battalion contingent with the prestigious sword in hand. The sword symbolized honour and chivalry, and therefore, the bearer was obligated to have an iron grip. A tremendous amount of respect was attached to this position of leading the battalion with a sword in my hand, and for that opportunity, I considered myself truly blessed!

With God's grace, the countless hours of practice were made sweeter when we could finally march without a single hitch, and all of us looked forward to the day when our parents would be seated in the spectator's gallery, cheering us on as we would march in front of Chetwode Hall.[10] I was particularly excited by the thought of my parents watching me lead from the front with the coveted sword shining in my hand, their eyes moistening as joy engulfed their souls. What a beautiful moment it would be; just the thought of it made me feel dizzy with excitement! But alas, Man proposes, God disposes! Fate was planning a different course for me altogether, although I am grateful to the supreme power for blessing me with the five-spoke golden star on my shoulders in the

---

10 Named after Field Marshal Philip Chetwode, the Chairman of the Military Committee which recommended establishing the Indian Military Academy, Dehradun, in the year 1932.

manner in which it materialized. It all started with our juvenile minds drinking beer; an incident that will forever be etched in my memory as one which caused a series of unfortunate events that were beyond my comprehension. We were barely out of our teens, and naturally, we all indulged in some good old mischief! It was close to the passing out day, and as such, our spirits were sky-high. Therefore, a douse of old-fashioned beer further fuelled our already high spirits. As per the rules, Gentlemen Cadets in their final term were issued one bottle of beer per head; that, too, only on certain days, referred to as the 'issue day'. However, while the protocol was that we could drink one bottle per head, the quota of non- drinkers was largely sufficient to raise the spirits of the hardcore and not so 'gentlemanly' types! So, on one such 'issue day' just before our passing out, our group decided to unwind at the Academy swimming pool while enjoying our chilled beers. This gradually descended into a poolside party. After our quota of one bottle was down and our throats were satiated, we felt the desire to indulge in some soul therapy and longed for some music without considering the fear of *facing* the music! Soon enough, thanks to one of our friends who hailed from Dehradun itself, the music was arranged. We were thoroughly enjoying our time and celebrating our last few days in the Academy before our passing out parade, which was just around the corner, when the festivities came to a screeching halt as a messenger walked in bearing the dreaded news of our Company[11] Commander's muster fall-in. Everyone immediately rushed to the company lines for the fall-in, except for me, since I had found the ambience of the pool too merry to leave.

Much later, having abandoned the fall-in, I reached my room without an inkling of what was in store for me. I was informed that Major Muthana, our Company Commander, was not very pleased with my absence from his fall-in. Reality struck the next morning, when I was standing outside the Company Commander's office ready to face the music, the antipathy to our soul therapy! My heart pounded inside my chest, as I stood there gazing at the brass logo of the Academy placed on a blood red and steel grey background right outside the office. I was called in. The terror intensified with every step that I took. Major Muthana was seated across his glass-top table and his icy stare seemed like the lull that precedes the storm. During that time, coincidentally, Dehradun was battered with torrential rains and streaks of lightning. The sound of the booming thunder rattling the windowpanes was supplanted by the holler of the Company Commander. 'Why were you absent from my fall-in?', he enquired. Blame it on me or the hangover after having one too many, my smug reply was that being a battalion appointment, I did not consider it compulsory to attend the company

---

11   A company is a subset of a battalion, with three platoons under every company. Each platoon comprises forty soldiers.

fall-in. Major Muthana, who was generally a calm and composed person, on hearing my reply, lost his temper and shouted, 'Do you know who made you the Battalion Under Officer?' In hindsight, I wish I had kept quiet, and things had not escalated to where they did. However, I instantly removed my appointment tabs that adorned my shoulders and tried handing them over to Major Muthana, saying, 'Sir, with all due respect, I'd rather be a nobody than be somebody because of someone.' The distressing silence made its presence known against the noise of the ticking clock. The sound of the chair scrapping against the floor broke the silence in the room. 'The tabs will be taken back but ceremoniously,' Major Muthana exclaimed before leaving the room.

The very next day, I was marched up on a charge sheet to the Deputy Commandant of the Academy. The Brigadier read out my charge written in black ink and asked me what I had to say in my defence. I decided to say nothing. I was informed that the offence committed by me was grave, and for my conduct, I should be withdrawn from the Academy; however, since I was to pass out in a couple of days, I was only being de-tabbed, for which I should consider myself very fortunate. I was marched out of the office of the Deputy Commandant, and even before Major Muthana could come out, I removed my tabs and tried handing them over to him. 'Keep them as souvenirs,' he said. To this day, those tabs which say 'Battalion Under Officer' are with me, and every time my wife opens the iron boxes muddled with dusty memories, she berates me for not discarding all the old stuff. Looking at the tabs kept wrapped in a cellophane cover deep inside the box, I always smile, and without a whisper, simply thank Major Muthana for letting me pass out with my course buddies! As I was no longer the Battalion Under Officer, as per procedure, I was not allowed to carry the coveted sword and lead my battalion contingent. In a split second, I went from a mountain to a valley in my heart. I was to be among the marching contingent holding the rifles. Since only the last few practice sessions were left before the actual parade, the Adjutant of the Academy, who was responsible for the conduct of the parade, issued orders against my inclusion at this stage as I had not practiced my march with the rifle, and one single step out of rhythm could put the entire parade in jeopardy.

I finally faced with the true enormity of the consequence of my monumental foolishness. And how? I was left with no option but to sit out in the stands and witness the passing out parade of my course as a spectator. One callous act had capsized my role in the parade: from the Battalion Under Officer to a bystander. Suddenly, the mere thought of me witnessing the parade alongside my parents jolted me. To dilute this humiliation, I walked up to the Academy post office and pushed out another telegram home, hoping that it would reach in time. The honey-coloured telegram paper had my note written in blue, which read, 'The Parade is cancelled, do not come.' Even after returning to my room, my mind was in constant turmoil, and I kept tossing and turning in my bed. Looking back, I cannot help but feel the hand of destiny interfering with my passing out and pipping ceremony. The mere thought of my parents not attending the parade bugged me immensely. Their absence meant that there would be nobody to put the brass on my shoulders. At the same time, I also felt guilty about sending the false telegram home. It was too late to amend it anyway, and nothing could change the fact that now I would be witnessing the passing out parade of my course from the sidelines. Until the previous day, I was to march with a sword in the parade, and today I was not allowed to thump my direct moulded sole (DMS) boots on the tar-laced Chetwode parade ground, the ground on which every Indian Army legend had set foot.

The last seventy-two hours I spent in the room that had been home to me for the last year were filled with nostalgia. I was all geared up to join the Indian Army as an Officer. I was grateful for the thick bond I had developed with my coursemates, something that I felt would last for eternity; and yet, I felt alone, incomplete and disheartened. This wasn't the climax that I had prayed for; in fact, it was the antithesis of how I thought my story would end at the Academy. My final hours were supposed to be spent welcoming my parents, watching my father beaming with pride as I became the first Army Officer in my family. Instead, here I was, attending protocol dinners with my course fellows and their families, but without my parents.

In a room swamped with people, I felt alone. As I went off to bed the night before the parade, I still wondered who would be pipping me. The next morning, I watched my coursemates donning the ceremonial uniform with excitement oozing through their eyes. They hurriedly left their rooms by five in the morning to participate in the parade, while I was still sitting on my bed wrapped in my green flannel blanket, sipping the tea brought in by our orderly, Roop Singh Negi. He had been working at the Indian Military Academy since the British era. His charter of duties involved primarily preparing the uniform of the cadets, as well as other minor administrative jobs. In his late fifties now, Roop Singh would often end up narrating incidents and anecdotes of the 'British era' and officers in those pre-independence times to us, which would most often be extremely informative and interesting to hear. Whenever Roop Singh was in the mood for such stories, all cadets would huddle around him listening to his, what for us were, fantastic tales. Almost every time, his stories would eventually end with a lengthy moral lecture for all of us. Indeed, Roop Singh was dedicated to his work and was a very popular figure among the Gentlemen Cadets. He was like a father figure to all of us. He was particularly fond of me as I would refer to him as 'Baba',[12] and many a times, he would bring kheer[13] for me, prepared

---

12   Father.
13   An Indian sweet.

by his wife, which I relished a lot. Baba was a staunch supporter of the British officers and would motivate all of us to follow in their footsteps by narrating incidents to illustrate their conduct. He would fondly recollect how, as a teenager, he was pulled out of a situation wherein he had almost reached his grave. Baba had lost his parents in a landslide in the upper reaches of Uttarakhand and was on the verge of ending his life as a result of extreme poverty and hunger. That was when he was spotted by a British Officer in the dense wilderness. The Officer not only provided him with food, but also brought him along to the Indian Military Academy. Baba was thus enrolled as an employee in the Academy and was witness to the complete transformation of the Academy from British control to the Indian flag hoisting on 15 August 1947. Despite the change at the helm of affairs in New Delhi, as per Baba, there was no change in the Academy curriculum. Baba would mention with pride that he had polished the boots of officers like Field Marshal Sam Hormusji Framji Jamshedji Manekshaw and many other very senior and prominent army officers. Even in his late fifties, Baba would conduct himself with enormous dignity, always taking tremendous pride in his job, even when it involved polishing the boots and preparing the uniforms of the Gentlemen Cadets.

Baba was a moving encyclopaedia about the history of the Indian Military Academy; he was an institution in himself. He could rattle out our course curriculum for the entire year in such a systematic sequence that we were left marvelling at his memory. He had lived the routine of our Academy year after year, all his life, and it showed. He could even predict the weather for our passing out parade, and in many cases, he could even foretell the arm/service a particular Gentleman Cadet was going to be commissioned into. He was *the* complete authority on the Academy and still carried its British lineage with utmost respect and dignity. On 9 June 1984, the D-Day for our passing out, noticing that I was still sipping my tea with no sign of hurry, Baba walked up to me. 'Beta, tum parade mein kyun nahi jaa rahe?' ['Son, why are you not going for your Parade?'], he asked me. Baba was concerned. 'Company commander ko yeh pasand nahi aayega' ['The Company Commander will not appreciate this'], he said. I didn't want to share my shameful incident with Baba at that moment so I simply replied that I would be going a little later. Baba's concern was like that of a father for his child, which brought out a peculiar emotion in me. I then decided that I would want to get pipped by Baba. I got ready, wore my newly stitched olive greens, and handed over the shining brass stars to Baba. There was a moment of silence as both of us looked at each other. I do not know whether it was the look on my face or whether it was the sheer experience of having witnessed so many parades and this time noticing something unusual, but all Baba asked me in his native tongue was, 'Aap ke ghar se koi nahin aa raha hai' ['Is nobody from your home coming?']

I did not have the words with which to answer him, so with a firm look I simply requested Baba to put the ranks on my shoulders. While I was dreaming about my proud parents pipping my shoulders with the well-earned stars amidst the glorious ceremony; yet now, I was getting pipped all alone in my room with no one with me other than Roop Singh. Baba had himself done the Brasso on the stars himself the night before, and now, as he held the stars in his hands, I remember his eyes moistening and his fingers trembling. With his hands that always reeked of Brasso, Baba pipped the stars on my shoulders. With a trembling voice, he confessed that in all his forty years at the Academy, this was the first time he had ever had the honour of giving the ranks to any Gentleman Cadet and making him an Officer of the Indian Army. As soon as he had finished securing the stars, I touched his feet, and the old man broke down as he touched my forehead with his age-spotted hands. No other words were exchanged between us as we both hugged. I could feel the warmth of that 'jaadu ki jhappi' ['the magical hug'], and the blessings radiating from Baba were enough to energize my entire soul and reduced my ocean-deep pain. I do not remember how long I continued in Baba's arms, but I do remember that those couple of minutes were transformative for me. I was mesmerized and felt the prowess of God Almighty blessing me with all the success in my future ventures. Tears rolled down my cheeks. I was at a loss for words and simply touched Baba's feet one final time before leaving the room.

I cherish the moment that I shared with Baba till date, and I believe that his genuine and heartfelt blessings played a significant part in me having an unblemished military career.

I have been in uniform for almost thirty-eight years now, and during these years, there have been many times when the odds were against me. But I always ended up counting on and remembering the blessings showered on me by Baba, which, in my opinion, ultimately helped me in overcoming all odds with my head held high. Moments in my life such as being surrounded by a rebel militia for three months without food in Africa, surviving a grenade injury with its splinter still embedded inside my body, and being declared dead in the operation theatre of the Army Research and Referral Hospital, New Delhi, are some of the momentous times when I could feel the warmth of Baba's hands on my head and his blessings paved the way in ensuring my survival. What finally remains for me is that in life we must remember to count our blessings. While it is okay to look skywards for the solace of God whenever we are in a fix, we must realize that he resides among us, in our elders and well- wishers. You never know when, where, and how God Almighty may come your way and match his footsteps with yours. We need to learn to be humble, grounded and respectful towards our fellow human beings; each one of us is God's creation and truly represents God in his purest form. Every religion in the world talks about the Almighty residing in our souls, and yet, we don't value the people around us. We have to prioritize respecting God's creations even more than visiting a temple, mosque, Church, or even a gurudwara. The world would be such a wonderful place to live in if each one of us remembers to respect our fellow human beings. You need not always look skywards for God Almighty; remember, he is amongst us and it is only a matter of feeling him, believing in him, and connecting with him. I assure you that your life will change forever the day you start looking for *God on earth*!

4

ADAPT AND STOP SMOTHERING CAPTAIN SMOKEY

While the penultimate days prior to getting commissioned had been extremely trying, my overall merit while passing out from the Academy remained high, and I was within the envious 'Super Block'.[14] Officers who were part of this esteemed

---

14 Super Block is the merit list of the top twenty Gentlemen Cadets of the passing-out course at the Indian Military Academy. They are allotted their choice of arm or service.

'Super Block' had the liberty of selecting the regiments of their liking that they wanted to get commissioned into. Being a first-generation officer and having no lineage whatsoever to any of the regiments, I was naturally confused. The fact that I had no one on my side to provide guidance or advice further complicated the matter. In any case, I ended up opting for the 'Mechanized Infantry,' and my choice in this matter was greatly influenced by the fact that the Mechanized Infantry was termed as the arm of the future and the division at Hisar which was just seventy kilometres from my hometown used to always have Mechanized Infantry Units under command. By God's grace, I happened to command the same Division almost three decades later. Indeed, such is Destiny! I ended up getting selected for my first choice and was accordingly directed to report to Hisar, though after a short leave. After getting commissioned, I immediately boarded the train to visit my parents at Rajgarh. I had not informed at home as I didn't get the time, owing to the action-packed last couple of days at the Academy. The train journey to my hometown seemed longer than ever, and the sound of the tea seller calling out to people was no longer annoying but was like music to my ears. The following day, I reached Rajgarh at around 4 p.m. That day, my hometown felt different. I had set foot at my place of birth as Second Lieutenant for the very first time. After deboarding, as I was walking towards the exit, I heard a faint holler. As soon as I turned, it was none other than Ram Kunwar ji who was rushing towards me. Ram Kunwar ji had been running a tea stall at Rajgarh railway station for as long as I can remember. He immediately recognized me; I wonder how as I was half my size while passing out from the Academy, owing to the rigorous physical training. I touched Ram ji's feet and it was truly like homecoming. 'Raju beta aa gaye! Iss baar kaafi samay baad aaye. Chai peeye bina toh nahi jane denge' ['Raju, Son, this time you've come after a long gap. I will not let you go without having tea'], Ram ji exclaimed. 'Bilkul Chacha, aapki chai peene ka mauka nahi chodunga' ['Absolutely Uncle, I will not miss the chance of having your tea'], I replied while resting on the wooden bench next to his stove. While the tea leaves were boiling, the whole town got wind of my arrival. While I enjoyed my cup of tea, which was filled with love, loads of ginger, and nostalgia, thousands of people thronged the platform for my welcome. I was the first commissioned Army Officer from Rajgarh tehsil, and hence, it was a matter of pride for everybody. In no time, I was standing on a pedal rickshaw with garlands around my neck and people walking alongside in huge numbers. After traversing the bylanes of Rajgarh, I reached a conspicuous blue gate, the entrance to my home. By then, my mother had been informed about my arrival, and there she was, standing at the doorway with her customary 'pooja ki thali' ['plate used for praying']. I thoroughly enjoyed the attention I garnered from my family and neighbours. That moment, I felt as though I had done something right in life. The very next day, I travelled to my village, Sulkhaniya Chota, to meet my grandfather, who has been my guiding light since forever. I am because he was. He was the foundation on which I placed all my life decisions. After a pleasant week amidst my loved ones, I was ready to leave for my first place of posting. I was excited yet nervous. I bid adieu to my mother, who was smiling when I was in front of her, but I knew she would have moist eyes the moment my back faced her. Such is the military life; separation is part of our duty. I have been away from home since the age of ten, but even then, every time I bid goodbye, my heart aches. My heart feels heavy and empty at the same time, and I don't even know how that works. So, after a well-earned leave post commissioning, I reached Hisar railway station on the stipulated date. Thankfully, Hisar was not very far from Rajgarh, and hence, I was familiar with the place. As is tradition, the reception of every newly commissioned Officer is done by his immediate senior; and Lieutenant Joshi, my senior subaltern, was at the station to welcome me. He surprised me by welcoming me with a garland in the traditional Indian style at Hisar railway station. A wonderful gesture indeed! The vehicle that had come to receive me was the iconic 'One Ton',[15] which used to be one of the primary vehicles among the transport fleet of a battalion in those days. The 'One Ton' vehicle was known for its supernatural drive, purely based on its mood and

---

15   Nissan Jonga-Jabalpur Ordnance and Guncarriage Assembly.

attitude! We still wonder—Is it the weather on a particular day or the mood of pthe driver, that would decide if the vehicle wished to move further or not? Till date, it is considered to be a miraculous invention, and a stunning innovation by a genius with no parallels in capability, despite the most modern fleet of vehicles. The icing on the cake used to be its astute appearance and demeanour which would put any world-class vehicle to shame!

Prior to our departure, Lieutenant Joshi had briefed me that the cantonment was approximately fifteen kilometres from the railway station. I was fine with the distance as I was to just sit and enjoy the ride. While matching my body movement with the jerks of the 'One Ton', the vehicle came to a screeching halt well short of the cantonment. I wondered what had happened, and suddenly, the driver turned the ignition off. I got off the vehicle. 'Punia, now it is a test of your strength, so kindly push the vehicle for the rest of the journey,' Lieutenant Joshi remarked. It was the month of June, and the temperature was sizzling at almost 50°C; and here I was, glancing in different directions with wide eyes. I thought it was a joke, but Lieutenant Joshi's stern face explained otherwise. Although I could not find any logic to his command, I had no option but to follow it. I started to push the vehicle with all my might as the driver and Joshi Sir remained seated in the vehicle without displaying an iota of emotion. As I huffed and puffed all along the route, the civilians watched me and were having a hearty laugh, and so I could not help but ask myself, 'Is this why I had been working so hard for the past eleven years? Is this the Army? People in Rajgarh were welcoming me with such pomp; if they see me like this, they'll laugh at themselves.' By the time we reached the unit location, my uniform was completely drenched with sweat, and I was at my wits' end due to exhaustion. 'Okay, Punia. Thanks for a smooth ride,' Lieutenant Joshi remarked while moving towards his room. I was stunned. 'I could have died,' my mind screamed, but I thought nobody had time to deal with what I felt. I was directed towards my designated room, which was the frosting on the cake; the so-called room had to accommodate six officers! While my morale took a hit looking at the dismal administrative arrangements in the battalion, I knew for a fact that they were a result of the Hisar Military Station being the newest military station in the country. It had recently been inaugurated and was still in the process of setting up accommodation and other infrastructure. Thus, only temporary structures had been created for the officers while the soldiers were still living in tents. While I could not do much to improve the situation, I was perturbed by the thought of what impression my parents and relatives would carry of the Indian Army, if, and when they planned to visit me, owing to the proximity. Even so, my days at Hisar were a total blur, with hectic training as well as administrative duties. Every day, our routine would involve a physical training parade at six in the morning, which was compulsory for everyone at the station. Thereafter, the battalion routine did not spare a single minute of rest till the evening, and finally came the games parade in the evening which used to culminate in the company roll call, which again was compulsory for a young officer like me. By the time one returned to their room, it was time for dinner.

The senior-most dining-in member in our Officers' Mess, Captain Duhan, expected me to check the menu for dinner, as well as inspect the Officers' Mess kitchen for its hygiene and sanitation prior to his arrival in the mess. I was sharing my room with Captain Duhan, also referred to as 'Smokey Duhan', who would ensure that I did not get more than five minutes to even take a bath post the games parade and the evening roll call as I was supposed to be in the mess to receive him. Once inside the mess, my everyday experience would be nothing short of traumatic. With 'Smokey Duhan' at the helm of things, on most days, the evening affairs would continue well past midnight. When the drinking session used to start, initially, there was an old-school British elegance in the air. However, by the time it was midnight, the general lingo being used by 'Smokey Duhan' was the choicest of Haryanvi,[16] with frequent use of unprintable profanities. During those evenings, a number of times, he would ask for dinner to be laid out, and accordingly, the mess Havildar[17] would smartly report by way of an announcement in the classic

---

16 The language spoken in Haryana.
17 Sergeant.

military style that the food was ready. But as soon as the report was given, Captain Duhan would invariably reply, 'The food needs to be warmed up!' This would repeat at least four to five more times, and with every such report, our stomachs would pray that hopefully this time he would agree to start the dinner. As bachelor members of the mess, and as is the protocol of basic mess etiquettes, we were forbidden from getting food in our rooms. Captain Duhan, during these evenings (till midnight and sometimes well beyond), would enjoy his drinks while conducting classes for all the young officers on mess etiquettes and basic protocols that were mandatory to be followed in the mess. While his long lectures and associated anecdotes went on, he would always be on the lookout for anyone who looked groggy or whose eyes were partially shut, showing signs of disinterest in his oratory skills. He would immediately send the defaulting officer for guard check and would then wait for his return before starting dinner, which, in turn, sometimes prolonged those evenings to the wee hours of the morning!

As the days went by, one could make out, just by looking at our haggard faces during the morning physical training fall-in, that we were suffering from an acute lack of sleep. And yet, 'Smokey Duhan' gave us no respite. Among the young officers, I, being the junior-most, used to dread the evening rituals in the mess; and there was good reason to it. By the time I generally returned to our room, it used to be two in the morning. Before I could settle down in my room, I used to come face to face with a board nicely prepared by the battalion 'Intelligence Section', which was on permanent display outside our common washroom. Lest I miss out, the board displayed the figure '0400' in bold and italics. My readers might be confused as to what this figure denotes. To put it simply, since all six youngsters were staying in a common room with a single washroom, I, being the junior-most, had to clear the washroom by four in the morning since after me, five more officers, as per ascending order of seniority, were to use it and be ready to attend the physical training parade at six in the morning. Imagine the effect this board used to have on me every night before I could hit the sack, knowing I had to be up and ready for the next day in about sixty minutes! But that was not the end of it; prior to me settling down in my bed, I had another major task which needed to be executed with caution as well as stealth. On my first day itself, my immediate senior, who, till then, had been the junior-most, happily handed over this sacred responsibility to me. My job was to slowly approach 'Smokey Duhan', who by that time had fallen asleep with a lit cigarette tucked precariously between his fingers and remove it from between his fingers without waking him up. Stealth was of utmost importance, as all hell would break loose if he woke up while I was in the act of removing the cigarette. Many a time, he used to wake up while I was on the job, and with a bottle of whisky carefully tucked alongside his stomach, he would howl,

'STOP SMOTHERING ME, YOU YOUNGSTER!' The words still ring in my ears to this day.

Difficult life situations teach you many things; on the one hand, you could be patient and have a high degree of endurance; on the other, you could learn the art of deflecting and avoiding such difficult situations. I was aware that this ordeal would be a daily affair, and so I started having my dinner immediately after the company roll call in the dining hall with the soldiers of my platoon,[18] and thereafter, I would quickly join everyone in the Officers' Mess. The second change I brought about in my routine was to get done with the washroom routine immediately after dinner, by which time Smokey Duhan would be so fast asleep that I could easily remove the cigarette from his finger. With these minor adjustments in my daily routine, I successfully improved my odds at surviving the 'Smokey trauma' and was able to ensure that I could get at least three hours of sleep every night. Today, when I look back at these distressful, and yet hilarious experiences, I cannot help but feel that such small occurrences and little incidents introduced me to the value of adaptability quite early in my professional career. Life is a mixed bag of happiness and sorrow, achievements and failures, and comfort and distress. It is therefore extremely important for all youngsters to be able to deflect thorny situations and adapt to those which cannot be warded off. I am now of the firm

---

18 Each platoon comprises forty soldiers.

belief that whatever nagging problems life throws at you, there is a viable solution to it; one just has to look for it and exploit the evolutionary advantages of adaptability. As for me, my education in those early years in the Army ensured that my instinct for adaptability became my biggest asset for the rest of my military career. Today, Smokey Duhan is retired, and he has quit smoking and drinking; yet, every time we meet, I definitely make it a point to say to his wife Alka, 'Hope you are not smothering him!' and we all share a hearty laugh at the expense of Smokey Duhan!

There is a single word, which, if learnt early on, can go a long way to ensure that you lead a comfortable and successful life, and that word is responsive. In today's world, when even traffic lights change their colour depending on the quantum of traffic, we need to be adaptive because there cannot be one solution to every situation throws our way. Life is all about taking the right decision at the right time, and that can only be feasible when we learn to be responsive. As a military man, I was required to deal with the lives of my soldiers, and even a fraction of a delay in making the right decision could cost me my comrades' lives; and hence, I am well aware of the importance of being responsive and making the right decision at the right opportunity. Resistance to change is natural as nobody is willing to move out of their comfort zone, which is why it is my opinion that we struggle with becoming responsive. However, there is no substitute to being responsive in life; the earlier you learn this hard fact, the better and happier your life would be. Be like the 'rounded pebbles' in a riverbed! The rounded pebble is the one which learns to change its shape without losing its stiffness, which is its basic trait. It only changes its shape in order to survive in the fast and furious water current. However, it does not compromise on its hardness, which happens to be its strength! The pebble which does not change as per the situation gets washed away.

I sincerely thank Smokey Duhan for teaching me, very early in life, the importance of being responsive.

5

RAJDOOT OUTSIDE THE BOX

In the Army, Units are generally posted at one location for a few years before being moved to another. After a few demanding, yet fruitful years at Hisar, our battalion moved to Patiala in the year 1986. Obviously, there was a lot of excitement in the air regarding our new location. Being a peace station, we all looked forward to the general comforts and good times waiting for us at Patiala. The married officers were especially excited since the mere idea of being able to stay with their families was enough to bring in the much-needed euphoria! We, the bachelors, on the other hand, had our own perception of a better quality of life, since the idea of being able to get a room to oneself was enough reason to celebrate. Probably the most significant, and yet, most anticipated reward of a peace station was the humble desire to have one's own private and exclusive bathroom.

Our Commanding Officer, Colonel K.K.D. Prasad, was of the firm view that for any Commanding Officer within a Unit or a Battalion in any peace station, the biggest challenge was the management of officers, especially the bachelor officers. Unlike today, when most youngsters in a Unit are either out on courses or they have moved out on postings as instructors or on staff appointments, in those days, the number of bachelor officers in every Unit used to be around eight to ten, depending on the location and operational responsibility. Considering this, one has no option but to agree with Colonel Prasad, since the number of bachelor officers was directly proportional to the number of incidents requiring their Commanding Officer's intervention! We were like overgrown naughty kids busy with our 'monkey business' all the time ('monkeys', in those days, had a plethora of business options)!

Unlike every other battalion location, where the unit campus is generally compact, Patiala was unique. While our battalion was located on the Patiala–Sangrur highway, our Officers' Mess was at a distance of over seven kilometres, in the heart of the city. As a result of our daily physical training and administrative routine, we bachelors found it to be extremely cumbersome to go to the Officers' Mess for our meals. To obviate the wastage of time due to the commute, we used to call for our meals at the battalion location in our designated living rooms. Everything was going well and as per our plans, until the news of us having developed the habit of refraining from using the Officers' Mess facilities for our meals reached the ears of the Commanding Officer. Colonel Prasad was extremely annoyed and passed strict orders for all bachelor officers, making it compulsory for them to have their meals in the Officers' Mess. The only exemption granted to us was for breakfast. The orders, in effect, meant travelling to and from the unit location to the Officers' Mess twice every day, once for lunch and once for dinner. Now, we, bachelors with very little service, had only just moved from Hisar, which being a field area, eluded us from the luxury of affording a two-wheeler. So, for this tiresome daily commute, we started using some of the battalion's vehicles on our own to travel to the mess for our meals. We had adequate measures in place to ensure that the fact that we were using these army vehicles for our private purpose remained as secret as an operational plan. However, we never catered or planned for the unique position the Commanding Officer holds within the battalion, from whom nothing goes unseen or unheard. The Commanding Officer's intelligence network within the unit can beat even the Inter-Services Intelligence (ISI) of Pakistan! Many a times, a detailed report is available with the Commanding Officer even before an individual act is accomplished! It was said between us that even God

Almighty, prior to writing the destiny of the individuals of a battalion, has to seek 'on file approval' of the institution known as the 'Commanding Officer'! Therefore, it was just a matter of time before Colonel Prasad found out about the bachelor officers using official vehicles twice a day to have their meals in the Officers' Mess. He was furious once he came to know about it, and so, to vent his anger, a quick officers' conference was summoned, in which the Commanding Officer with his roaring voice amply justified his code word 'Tiger'! and passed a strict directive that come what may, no officer could ever misuse official service transport for personal usage again.

With no official vehicles to fall back on anymore, the erstwhile cumbersome task of commuting between our Unit location and the Officers' Mess quickly spiralled into a herculean exercise. The challenge used to augment itself when we had to move during the night, and there was a good reason for it. All this was taking place in the late 1980s, which happened to be an era when the entirety of Punjab was enduring a phase of militancy post Operation 'Blue Star'. Therefore, one could hardly find any public transport on the road after sunset; and to add to our misery, the stretch of the highway between our battalion location and the city was completely uncivilized, with dense forests on both sides of the highway. We therefore decided to present our very valid concerns against the Battalion Commander's orders to the Second-in-Command of the battalion. Major Prakash Arora instead justified his codeword, which happens to be 'Lamb' for every Second-in- Command of the Indian Army! With a dry smile, he simply reinforced the Commanding Officer's directive. Thus, there we were, standing by the highway, trying to hitch hike and somehow reach the Officers' Mess. Mercifully, most of the time, on our way back after dinner, we would manage to get a lift in the Duty Officer's one ton vehicle, which was being used for a night-guard check by him. While at the Officers' Mess bar, over drinks, we would crib against this order to the best of our abilities and make new plans every day as to how we could put an end to this daily ordeal of an 'un-officer-like commute'. With the following sunrise, those whisky-induced carefully crafted plans would simply be forgotten! Bicycles were something we could afford but cycling on the highway at night was not recommended, as it was obviously unsafe, and even more so as the highway used to be pitch dark, with no streetlights installed. Every day, our discussions used to culminate on the possibility of applying for a bank loan to buy a motorcycle, but these discussions could never fructify into any sort of viable action. The readers must realize that unlike today, in those days, our salaries were not as handsome as what any young officer of today gets immediately on commissioning. To add to it, the reason I still have no explanation was that we, the young officers, used to compete among ourselves as to who could generate the maximum monthly mess bill so that he could be crowned the 'Royal Officer' of the month. In this Faustian bargain, all our mess bills ended up being quite hefty, and at the end of the month, our salary accounts had no reason to smile!

Once, as per our daily evening ritual, Prakash, Kishor and I were again standing by the roadside, looking for a lift to reach our Officers' Mess. The wait that day was, for some reason, longer than usual, as there were hardly any vehicles on the road. It was pitch dark, and we were wondering whether we should keep waiting in the hope of any suitable vehicle or simply forego the mere thought of dinner. Finally, in the darkness, I could spot the reflection of a single headlight approaching us from a distance. We were aware that it must be a two-wheeler, and we happened to be three, but something was better than nothing, and we started to desperately wave our hands and shout for the motorcyclist to stop. Sensing us to be in desperate need of help, the motorcyclist stopped a couple of metres ahead of us. It was dark, and initially we could not make out much, except the outline of the motorcyclist. We moved ahead and approached him. He was a burly Khalsa[19] on a Rajdoot[20] motorcycle with massive milk containers hanging by its sides. For readers who are not familiar with the rustic way of life in small towns and villages, even today, one of the primary means of transporting milk to households is through motorcycles. Also, unlike today, where there is a plethora of choices when it comes to motorcycles, in those

---

19 A Sikh gentleman.
20 An old motorcycle brand in India.

days, the two decent and sturdy motorcycles available and generally subscribed to were the Bullet[21] and Rajdoot. Being the senior-most among us, I signalled Kishor to mount the motorcycle before the Khalsa could have any second thoughts about giving us a lift. While Kishor mounted the bike, the Khalsa, without raising an eyebrow, looked towards Prakash and offered him the remaining space behind Kishor. Once both Kishor and Prakash were firmly seated, I patted Khalsa on the back, denoting my gratitude, and asked them to proceed. Yet, the motorcycle remained static. So, I again told the Khalsa to carry on as I would take a lift from the next vehicle which would pass by. Surprisingly, the Khalsa disagreed, and what he did next will always remain etched in my memory, partly because of the amusement associated with it and partly due to our heartfelt gratitude for this gentleman. The Khalsa began to wiggle his backside forward, and in the process, instructed Kishor and Prakash to do the same. Within a minute, the gentleman had almost crowned himself on the petrol tank of the motorcycle, with the two officers following him to the edge of the front seat; and miraculously, some space was now available for me to mount the bike. The Khalsa would not take no for an answer, and finally, in spite of all my protestations, convinced me to mount the motorcycle. Once the motorcycle started off, we had no idea how the

---

21   A motorcycle manufactured by the Royal Enfield motorcycle company in India.

Khalsa was managing to drive, although it must have been a sight to see three young officers precariously hanging on to those milk containers heading for their Officers' Mess. Initially, we attempted to give him directions but realized soon the futility of the exercise, as he was already aware of the location of our Officers' Mess in the Doctors' Colony. Through the course of our conversation with him, we realized that the Khalsa happened to be the milk supplier to our Officers' Mess.

We begged him to stop at least a hundred metres short of the Mess gate, so as to avoid the glare of the sentry stationed outside. But sure enough, the large-hearted Khalsa had to stop his hovercraft (Rajdoot) right in front of the mess gate, and the sentry saluted us with a smile on his face that conveyed a thousand words. Reluctantly, we were left with no option but to say, 'Jai Hind'. Quite a bit ashamed with ourselves, all three of us, with our eyes firmly gripped to the ground, quickly entered the Officers' Mess. The event had a profound effect on our post-drinks discussion that evening, and we passed a resolution to head for the nearest two-wheeler showroom the very next morning. The next day, the same sentry smartly saluted three officers of his battalion with enormous pride when they arrived at the mess on their sprawling new motorcycles! The smile on the sentry that day conveyed a different tale, but this time the curved lip displayed dignity and honour! The next morning, we ensured that the motorcycles would be in a single file and in perfect rhythm when we entered the 'Battalion Gate', and so, as we approached our physical training parade ground one by one, we could feel hundreds of eyes on us; the self-esteem and pride that all three of us felt cannot be conveyed in words. Later in the day, as I was walking past the office of the Commanding Officer, I overheard a discussion about the three of us taking place between Colonel Prasad and his Second-in-Command. Colonel Prasad sounded very satisfied and amused while telling the Second-in-Command, 'Now you understand my logic of making meals compulsory in the Officers' Mess, otherwise these young brats would have never been riding such smart glossy looking motorbikes!' Many years ago, when the order for compulsory attendance for meals at the Officers' Mess was passed by the Commanding Officer, we felt it to be disturbing and intrusive; and yet, today in hindsight, I realize and understand the logic behind the order of our Commanding Officer Colonel K.K.D. Prasad. While in many ways it was wrong for a Commanding Officer to dictate the kind of lifestyle that an officer should lead, it is an art, known only by a select few, to institute such rules which have a direct effect on the general routine, and which subsequently elevates the living standard of a young officer. And so, today I endorse the same order with all my heart and sentiments that order, which earlier made us youngsters uncomfortable. Now I can vouch with absolute conviction that the solution to most problems thrown at us by our environment lies outside the box, needing just that uncomfortable push towards it.

The phrase 'out-of-the-box solution' has a deep meaning which, when understood, can solve all our problems in life. Most of us make the singular mistake of focusing on or paying too much attention to the problem, thereby increasing the value of the problem manifold. However, the real solution lies outside the problem, which we in the Army refer to as the 'out-of-the-box solution'. Life cannot be a straight line. If the lines in an echocardiogram were straight, that is the end of life! Where there is life, not only is there is bound to be a problem, but there also has to be a silver lining at the end of the tunnel, meaning that there are definitely better times ahead of the problem. Therefore, don't be disheartened with the curves of life, as it is absolutely normal to move up and down, without which there is no meaning to it all. Do not get disturbed with these oscillations of life; they are part and parcel of life. Learn to take it easy and relax, but at the same time, remember to work hard, play hard, and party harder!

6

HURT YOUR EGO TO PROGRESS IN LIFE

By now, a few months had passed since our battalion had been established at the Patiala Military Station, and it was time for the sports season in the station to commence. The first sporting event that came up for competition was the inter-battalion volleyball tournament. Sporting events and inter-battalion tournaments assume great importance in the Indian Army; and in our case, this was a perfect opportunity to announce our arrival in the new military station by way of winning a sporting event. At the change of command that had taken place immediately on arrival of the battalion in Patiala, Colonel Jaswal had taken over the reins of the Unit, and he happened to be a diehard sports fanatic. He had made it very clear to all of us that sports were one of his primary focus areas, and had passed specific orders to the battalion that the volleyball team must focus on improving their game as losing was out of the question and would not be accepted. Thereafter, all playing members of the volleyball team were excused from their fatigue duties, and the only task that they were expected to do was to play volleyball. They were also provided with a special nutritious diet. The Commanding Officer wanted a progress report of volleyball training almost every day, and so, Havildar[22] Chattar, who was the senior-most playing member and was the in-charge of the team, was tasked with briefing our Commanding Officer on the progress of the team. Chattar was a senior Havildar in the Battalion and was soon going to be promoted to a Naib Subedar.[23] Therefore, apart from the obvious responsibility of the performance of the team, the personal stakes for Chattar were also extremely high. For the team, this meant that he would go the extra mile, and the gruelling schedule of the playing members ensured that the team was shaping up well.

While the team was practising well, to gain a better understanding of the nuances of the competition, Chattar

---

22 Sergeant.
23 Junior Commissioned Officer (JCO).

was interacting regularly with the team members of the neighbouring battalion since they had spent a considerable amount of time in Patiala. He was surprised to see the other teams being headed by young officers who were regularly practicing with their team. On further enquiry, Chattar was alarmed to find out that it was compulsory to have an officer as part of the playing six in the team. This rule was not in vogue in our previous station, and therefore, came as a shock since none of our officers were good at volleyball and could in no way qualify to be part of the battalion team. The new rule struck us all like a lightning bolt and created a crisis, considering the fact that only a few days were left for the tournament to commence. Colonel Jaswal immediately called for a conference in his office, which was attended by the Second-in-Command, the Subedar Major (the senior-most junior Commissioned Officer) and Chattar. Eventually, I was called in by the commanding officer and informed that I was the lucky officer who had been selected to lead the battalion Volleyball team, and he instructed me to join the team forthwith. Colonel Jaswal was pretty well known for his short temper, and therefore, I preferred to keep quiet. However, I aired my reservations to the Second-in-Command once we were finished with the conference. I explained to Major Arora that in all my life I had never played this particular game, and it would be unfair on my part to be the cause of defeat for the battalion. Major Arora did what he was known for, and simply smiled! The very next morning, reluctantly, I reported to the battalion volleyball court, only to find the Commanding Officer already present at the ground. He told me he was aware of my reservations, but also said that he had seen me play basketball, and since I was good at that sport, along with the advantage of my good height, it would not be difficult for me to pick up volleyball in this short period. I tried to explain to him that I had never been inside a volleyball court, let alone played the game, and it would be extremely difficult for me to play as an effective member of the team. To this, I only received a cold stare, followed by, 'My word is final and no logic on this planet can supersede that.'

With this, the Commanding Officer simply walked off, leaving me stranded with the battalion volleyball team. So, for the first time in my life, I entered a volleyball court and held a volleyball, which felt very strange at first as I realized that it was a lot smaller when compared to a basketball! My practice session began from that very moment and continued every day for at least six hours a day! Very soon, my life turned miserable, as once inside the court, no one treated me as special. Rather, Chattar was particularly harsh and would never lose an opportunity to remind me that I was the weakest link in the entire chain and would most likely be the cause of the battalion's loss. Chattar would ridicule me in front of the entire team and would spare no opportunity to humiliate me; and on every such occasion, I felt like walking up to Colonel Jaswal and requesting him once again to earmark another officer. But knowing our

Commanding Officer's temperament, I abandoned such notions and would simply continue with my practice. On most days, while Chattar and the rest of the team members would be giving me a disapproving stare, I would often want to scream and tell them that they had been volleyball players all their lives and had been practising for the last six months, in particular. How could they equate themselves with me? Yet, I had no option but to lump the humiliation quietly. The gruelling practice sessions continued, and yet, somehow, despite the best of my intentions and efforts, I was never in sync with the rhythm of the rest of the players.

On some occasions, during our evening practice, the Tiger (the Commanding Officer is often referred to as 'Tiger') would come to oversee our practice, and at the end of it, after praising the efforts of the team, he would single me out and tell me specifically not to let the battalion down. It is hard to describe the tremendous anxiety I was suffering from at that time. The pressure was getting to me, and this brought down the standard of my game even further. I kept on trying my best, and within the first week, hurt all of my fingers. Very soon, my evening routine after the daily practice session would be to quietly sit down in my room and treat my injured fingers with ice; but I had no remedy for my *ego*, which was getting hurt the most. The days flew past, and in no time, the tournament had already started. In the Division, a total of twenty teams were participating, and luckily, we were placed in a comparatively easier pool for our initial matches. Our boys were very proficient in their game, and soon we reached the finals of the tournament. This, in essence, meant that our team was amongst the best two teams in the entire Division. I had assumed that the 'Tiger' would be thoroughly impressed with our efforts, but rather than being pleased, the Tiger's hunger for victory enhanced manifold as soon as we made it to the finals. On the evening prior to our final match, the Tiger met the entire team with the aim to motivate us. I was not sure about the team, but stress levels only increased post his 'pep' talk. On the day of the match, as both teams lined up in the central volleyball court, the atmosphere around us was electric, with the Divisional Commander, who was a two-star General, in attendance to witness the finals. The referee lined up both the teams on the ground, and thereafter, took the permission of the General Officer to commence the match. As we started to play, every time the opponent team was to serve, Chattar would loudly instruct all players to cover me so that I did not get to touch the ball. I found this to be odd, and even humiliating, as this was something Chattar had never done in the previous matches. But that day being the all-important match, he probably did not wish to take any chances with my skills. It continued to happen again and again as the match progressed, with the entire Division watching. It made me feel extremely small, as if I was simply to stand in the middle of the ground and not touch the ball. Under those circumstances, I imagined myself to be the laughingstock of the event. After it kept happening a couple more times, I could not take it anymore and I simply walked out of the ground while the match was still on.

Colonel Jaswal, along with the others from my battalion, shouted at me to get back into the court; but by then, my ego was bruised beyond repair, and I had taken a conscious call to walk out. Needless to say, our team lost the finals since they had to play with one less player. Tiger was extremely annoyed, and after the match, he admonished me in the strongest possible manner, using profanities to the best of his vocabulary, as I stood in pin-drop silence. After this incident, I had requested for one month leave, and probably because of my visibly low morale, it was sanctioned. However, rather than going back to my hometown, I decided to spend the time in-station and utilize it to learn volleyball at the Netaji Subash Chander Institute of Sports in Patiala. At the institute, luckily for me, there was a volleyball coach who hailed from my region, and therefore, he took special interest in me. He spent all his time and energy into shaping me into a proper volleyball player. I had good height and the coach encouraged me to spike the ball into the opposite court with full force. As I was picking up the game quite proficiently. In my spare time, I continued practicing at the institute even after my leave was over. I was now desperately waiting for next year's volleyball tournament. Soon, it was time for the next volleyball championship, and the set up for the finals was exactly the same as the previous year, with one small deviation; this time, Havildar Chattar, rather than shouting for the players to cover me, was busy instructing them to make the ball for me so that I could spike it into the opponent's court! With every successful spike, I could hear a loud cheer, and Chattar would publicly compliment me for the same. Suffice to say, our battalion won the championship, and I was awarded with the player of the tournament award. Perhaps bigger than the award for me was the emotional hug I received from the Tiger after the match which I very fondly remember till date! I also remember Tiger driving with me firmly seated in the front seat, and the moment we reached the Tiger's den, he asked his wife to treat me with some coffee and cookies! My intent of writing about this incident is simply to point out that most of us consider our ego to be a negative influence. However, I beg to differ, and I am of the opinion that a human being without their ego is pointless. Ego is your self-respect, your pride and your honour, and at no stage in life must one compromise these. If your ego gets hurt, don't get perturbed; instead, rise and shine like never before. In fact, it is my sincere recommendation that, at times, 'Do hurt your ego to progress in life!'

7

THE CURIOUS CASE OF THE FOOT

As an Army Officer, one has to undergo a number of professional courses. Driving and maintenance used to be one of the courses which was conducted at Ahmednagar, Maharashtra. I was detailed to attend this course, and I remember working extremely hard during the course. Luckily, with the grace of God, I ended up topping the course. As a result of my outstanding performance, I was appointed the Technical Officer in the battalion by our Commanding Officer. My charter included ensuring the serviceability of the complete equipment inventory, including all vehicles, of the battalion. The primary concern for the battalion and the Commanding Officer was that the Infantry Combat Vehicles had to be in an operational state. These vehicles were of Russian origin and had newly been inducted into the Indian Army. The day I took over as the Technical Officer of the battalion, a Russian team had landed in the battalion to help us out in the maintenance of our fleet of combat vehicles during the initial days. Being the Technical Officer, I closely interacted with these Russian officers, and I had a wonderful experience. In spite of both sides not knowing each other's language, we ended up communicating quite well, even without the use of an interpreter. For some reason, I found the Russians to be in a constant state of happiness, they were quite jovial and cheerful. Their day would usually start with vodka, so much so that I remember instructing the wine in-charge of our Officers' Mess to keep the bar open even during breakfast hours! However, while at work, they were a bunch of thorough professionals, able to strip open the engine of an Infantry Combat Vehicle in no time. Their mechanical proficiency would make any world-class mechanic envious, and personally for me, it was a period of great learning, which held me in good stead professionally for many years to come. After initial trials by the Russians in my Unit location, I was detailed to accompany them to the Central Armoured Fighting Vehicle Depot in Kirkee, Pune. I was happy to accompany them, since besides the knowledge I acquired, the charm of being in Pune for a couple of months had its own attraction! While at Pune, I regularly started going for a game of basketball in the evenings at the Bombay Engineering Group, Kirkee, with a Russian Officer for company. The Engineering Group was responsible for preparing the Southern Command team for the Army Championship, and since the Russian Officer and I were proficient basketball players, the Bengal Engineering team was more than happy to accommodate us in the game.

This evening ritual continued for a month until one fine evening, when I twisted my ankle. By night, my entire foot was swollen. The next morning, I visited the doctor, and was advised hot water treatment and given an ointment to be applied on my foot. I still remember the name of that ointment—Medicreme. Back in my room, I somehow reversed the process of treatment, and would first massage my foot with the ointment, and thereafter, dip my injured foot into a bucket full of boiling hot water. And in my desperation to get back to the basketball court at the earliest, I continued this inverted treatment about four to five times in a day for a week or so. The effect was that the colour of my foot started changing to black, and I was naïve enough to assume that black was a sign of healing. It was much later that I realized what was actually happening to my foot. Medicreme, which I was applying on my foot frequently, was getting fermented as a result of the hot water, and in the process, the entire skin of my foot was getting burnt. Thus, the swelling on my foot, instead of reducing, multiplied many times, and I finally landed in the Military Hospital, Kirkee, with a serious case of 'burnt foot swelling'. The doctors told me that my entire skin was burnt, and the foot was swollen as a result of pus formation underneath. So, what was just an ankle sprain had now turned into severe skin eczema, resulting in complications requiring a strong doze of penicillin every six hours. I had become the laughing stock of the hospital, as every time the Commandant of the Hospital would visit on his usual round, the doctor attending on me would deliberately and elaborately explain to him exactly how I had burnt my own foot! I was put on very heavy antibiotics; yet somehow, the pus, instead of drying up, started spreading into my body. By then, I had spent almost two months without any improvement, and the worst part was that I was bedridden throughout, having been strictly forbidden from even walking a few steps. Jyotsna, a charming young Nursing Officer, had been specially deputed to take care of me, but of late, as a result of my frequent calls, she had even stopped responding to the bell. In fact, she would warn me not to misuse the bedside bell specially designed for me since I was a bed-bound patient. I was in deep pain; not only due to my foot, but now even both my hands and backside were swollen due to frequent injections. Seeing my foot, nobody would have believed that it had all started with a sprain, and now my condition had deteriorated to such an extent that the bedsheet under my foot had to be changed a number of times a day due to the pus oozing out.

Finally, one day, the doctor told me that having tried every possible treatment, the only option left was amputation since the pus was slowly spreading into the upper part of my body. Note that here I am talking about 1987, a time in the Army when mine blast victims were being evacuated from Sri Lanka (Indian Peace-Keeping Force) to the Military Hospital, Kirkee, and amputations were routine. Unfortunately, I had never heard the word 'amputation' before and was blissfully unaware of what was going to happen the next day. I had assumed amputation to be another type of medication which the doctor wanted to try on me since I was not responding to the earlier treatment. The next morning, I was completely amazed to see Jyotsna, with flowers in her hands, walking into the Officers' Ward and handing them me! I blinked a couple of times to confirm that it was not a dream and I immediately asked her what the matter was, and that is when she explained to me what the real meaning of amputation was. I remember my reaction to this new information clearly; I immediately started screaming and was uncontrollable. Due to the ruckus that I was creating, the Commandant was summoned, and still my theatrics, coupled with the choicest of abuses, only got worse. I was immediately discharged on disciplinary grounds, with a threat that the Military Hospital was going to write a complaint against me to my Commanding Officer. By the time I got discharged from the hospital, it was evening, and wearing a slipper in one foot and a shoe on the other, I boarded a general compartment of the Jhelum Express to head for Ambala, from where I could reach my Unit at Patiala. Luckily, the Jhelum Express started from Pune itself; therefore, I could manage a window seat with my injured foot nicely placed on the opposite seat. The co-passengers, seeing me wiping the pus oozing out of my foot, sympathized with me, and on being asked a number of times, I finally narrated the entire ordeal from the beginning right up to the moment I was about to be put on a trolley stretcher to be rolled into the operation theatre for my foot to be chopped off. One co-passenger advised me to get down at Kopergaon railway station and then go to Shirdi for the darshan[24] of Sai Baba, assuring me that with his blessings, I would be all right.

Till then, I had never heard of Shirdi Sai Baba in my life, but when you are suffering and at your wits' end, every piece of advice, even from strangers, appears to be divine guidance from the Almighty. Sure enough, I got down at Kopergaon railway station, and since it was night time, I had to spend the night at the railway platform. In the morning, I took a tonga[25] to reach Shirdi. Unlike today,

---

24 Paying tribute.
25 A horse cart.

in those days, you could walk straight into the temple without waiting in an exhaustive long queue. Inside the temple, I spent a couple of hours seated in front of Sai Baba without any rush or disturbance. Finally, once I had shared my sentiments with Baba to my heart's content, I got up, took a horse cart, and headed back to the railway station to catch the next train for Delhi. By the time I reached Delhi, my plan had completely changed, and instead of heading for my Unit at Patiala, I now boarded a train for my hometown in Rajasthan. I am surprised and amazed till today, wondering how my plan changed. Was it the intervention of some divine power or my inner self? Whatever the reason may have been, I am convinced that this was the turning point for my recovery. An officer, come what may, cannot go home without leave. However, I went, and I am grateful to Sai Baba for sending me home, otherwise I would probably have landed in another Military Hospital. My parents were shocked to see the condition of my foot, and upon hearing the entire story of my ordeal, my mother started crying. The next day my father took me to the local Ayurveda Centre and my treatment started the same day. I still remember that the doctor, Shri Mahavir Prasad (Ayurveda doctor), told me upfront, 'Son, you are in a very critical stage,' and that the poison had by then spread throughout my body. The moment he held my hand to check my pulse, I could feel the healing energy radiating into my body through the touch of his palm. This amazing experience of a healing touch was divine in nature, and for me, Shri Mahavir Prasad, at that particular moment, was probably the true incarnation of Sai Baba. The doctor explained to me that only poison can treat poison and that my blood had to be purified, so he immediately put me on medicine extracted out of a poisonous plant, with strict instructions to not eat anything, and instead, only survive on milk. I recollect going without food for a month, but luckily for me, within a month's time, I was in a condition to head back to Patiala. However, it probably took me one year to return to the basketball ground, which had always been my first love.

In those days, there used be almost no means of communication, unlike today, and therefore, in a way, I was absent without leave for almost a month while undergoing treatment at my hometown. Considering this, I would take this opportunity to thank Colonel Jaswal from the core of my heart for being kind enough to understand the circumstances for my absence without leave. It was only he who managed to spare me from official disciplinary action for my unauthorized absence. So many years have passed since the unfortunate case of my 'burnt foot swelling'; although with the passing of years the marks on my foot may have faded, not a single year has passed when I miss my darshan of Shirdi Sai Baba. Every time as I walk into the temple, I cannot help but steal a quick glance at my foot, which I believe owes its survival to Baba. Time teaches you a lot in life, but what you learn during difficult times remains etched in you forever. Sai Baba has followers from every religion in the world, and without his divine intervention, a tiny village with a few hutments could not have transformed into a 'smart city', which can even take pride in being home to an international airport and some of the best hotels in the country! Sai Baba's own life had been a struggle, but today he is blessing lakhs of followers every day. I am a firm believer in the fact that there is a celestial or a supernatural power which is controlling this world. What I am trying to convey from these life experiences is that you may call him God, Allah, Father, Satnam Waheguru ji, or by whatever name, but there definitely is somebody controlling this world. The best example of God's creation is humanity; after designing the frame with all its integral parts, God finally blessed the mortal with the *subconscious,* which truly is the incarnation of the supernatural. Try and connect with your *subconscious*, and you would connect with the Almighty. Learn to communicate with your *inner self*, and the Divine Supreme will guide you at every critical moment in your life. Let the Divine Supreme decide for you what is right and what is wrong. That is why I always say that killing a human being is a crime in any country, but the moment that same act is executed on the border, it's a matter of rewarding the individual with medals. So, who decides what is right and what is wrong? Your *inner self* does. God Almighty guides you at every moment in life and will only cease to do so once you stop listening to your *inner self*. Therefore, I always say 'hasten slowly' and listen to your subconscious, which is the real God within you, and the day you realize this, I can assure you that all your problems and worries would cease to exist.

8
FLAG HOISTING ON A SAND DUNE

In the good old days, the annual field firing by all Mechanized Infantry Battalions used to be the most important event of the year, in which the entire battalion used to participate religiously. Every individual was recalled from leave, and come what may, one could not miss this all-important activity. The only firing range available for carrying out the field firing used to be at Pokhran in Rajasthan. The exercise involved displacing all the soldiers and heavy equipment from their permanent unit location to the Pokhran ranges. Therefore, the primary challenge was to reach Pokhran rather than the field firing itself. Back then, the Indian railways was divided into various sectors based on broad gauge, metre gauge, and narrow gauge. The move of our battalion from Patiala to Pokhran was like a 'coup de main' since it required a changeover from broad gauge to metre gauge, and vice versa. So, a routine schedule, which would involve one week of firing at Pokhran, used to take a minimum of at least three months of train travel; that, too, if one was lucky enough to get the designated railway rolling stock in time at Suratgarh. The train itself used to be a mix of a variety of rolling stocks like open carriers for vehicles, covered wagons for the ammunition, and passenger coaches for the personnel, as well as a special first-class coach for the officers. The first-class coach used to be modified into a proper living room with special lights fitted and beds laid out, giving a run for its money to any five-star hotel suite today! An army field telephone was connected with the engine driver as well as with the train guard. The engine driver would seek the Commanding Officer's permission before starting the train. In fact, every time the train started, the railway guard on the train would ask for our preference for the next halt since for every meal we would take a break at some railway station enroute. At every such halt the Officers' Mess used to be laid out on one of the disused platforms, and in case of it being a dinner halt, the bar layout was customary. At every such halt, the station master of the station where we halted, along with the train guard, would join the officers on the train for a drink, and any further movement of the train would invariably get delayed since the station master (after a couple of drinks) would insist on everyone having food from his home! More than the food, it was the warmth and hospitality displayed by the station masters and their families, which used to be so touching.

During one such field firing movement, the Officer Commanding of the train, Captain Khatri, insisted on a long halt at Jodhpur so that all the officers could go for a short sightseeing visit to the Jodhpur Fort. Since the train could not be placed on an unloading ramp due to technical reasons, a Jonga was pushed down the railway bogie by the collective efforts of our boys, and it ultimately facilitated our visit to the Jodhpur Fort. There are numerous interesting anecdotes of such kind about these special military movements. One such interesting incident happened at Suratgarh railway station on our way back to Patiala post the successful completion of the field firing. Suratgarh used to be the transit halt and the place where the gauge change of the train used to take place. At Suratgarh, we would usually end up spending around a month waiting for our rolling stock. During this transit halt, a proper military camp was established next to the railway station. Our Unit's raising day was approaching, and we all had to reach Patiala at the earliest possible date to have the auspicious ceremony celebrated in the required manner. I was tasked by my Company Commander, Major

Baruah, to liaise with the Station Master for the early placement of the rolling stock since I happened to be a native of Rajasthan. The very next day, dressed in my smart black dungarees,[26] I walked up to the Station Master. A gentleman by the name of Meena was the station master. In the hope of getting some preferential treatment, I requested him in my local Rajasthani[27] to place our train in the earliest possible time frame. The response to my request from Meena was highly disappointing! Contrary to my expectations, he took out a shabby-looking dusty register, flipped through its pages, and informed me as dryly as possible that it would be placed in its due time, and was in fact likely to take some more time, as there were two more military Units in waiting, preceding us by a month. I returned to our camp location, thoroughly dejected and informed the same to my Company Commander. Major Baruah did not say anything, but his silence said it all. Once the news percolated down to the soldiers that the train was unlikely to be placed in time, and therefore, the company would not be able to reach Patiala in time to celebrate Raising Day, everyone's morale was a little low. That night, while I lay awake on my bed pondering over my failure and devising plans to convince Meena, my mind drifted to my childhood. I would sit for hours with my grandfather and hear stories about people and how times were changing.

---

26 Overalls.
27 Dialect of Rajasthan.

I remember my grandfather often telling me that we are living in the Kalyug (age of darkness) era and how everyone is living to fulfil their self-seeking desires. It is very difficult to get anyone to work. People work only when their ego is boosted and garlands of praises are presented to them. It is for you to judge whether the person in front needs to hear a few nice words of praise in order to get your work done by them. These teachings of my grandfather instantly gave me an ingenious idea.

The Raising Day was just a couple of days away, and by the next morning, I had formulated the perfect plan in my mind. Armed with this plan, I walked into Major Baruah's office tent, shared it with him, and got his in-principal approval. His message was simple—'Do anything you have to do, but the train must be placed on time.' Our battalion had a tradition of flag hoisting on Independence Day, which was also fast-approaching. So as part of my plan, I walked towards the Station Master's office with an invitation card in my hand to formally request him to be our chief guest for the flag hoisting ceremony. Once I handed over the card, the change in demeanour that Meena exhibited was nothing short of bizarre. I explained to him that all Army Units have this tradition of flag hoisting, and since the Unit at that time was camping within the geographical boundary of the railway station, he, being the Station Master, would have to do the hoisting of the national flag. Having heard what I said, he jumped from his chair with a huge smile on his face, shook my hand, and said, 'It would be a matter of honour, Sir.' On the morning of Independence Day, Meena turned up in his smart white uniform, complete with his railway peak cap, and hoisted the national flag on top of a sand dune with all the soldiers of my company singing the national anthem. This was followed by a cup of tea and distribution of sweets to the railway employees, who I had gathered so that they could witness this spectacle. The Station Master was beaming with pride, and it was clearly apparent that the experience of hoisting the national flag for him was emotionally overwhelming. While unfurling the flag, I could clearly see his watery eyes, and while delivering the speech post the flag hoisting, his voice choked as he felt overwhelmed. The white uniform with its golden brass buttons worn by Meena was a testimony of the immense importance and gravity that he bestowed upon the event. He had displayed extra initiative by mustering a local photographer who was also tasked to share the flag hoisting photograph with a local journalist! At the end of the ceremony, prior to his departure, he once again shook hands with me and requested me to visit him in his office in the evening.

At his office, he had organized a special tea for me. While sipping the tea, without mincing any words, he simply asked me what he could do for me. Considering this to be my opportune moment, I explained to him the tradition and the importance of Raising Day for our battalion, and how the non-availability of a train would hamper the Raising Day celebrations. Meena simply smiled and asked me if the placement of a train the next morning would be all right. This time, it was me who jumped from my seat and shook hands with him and thanked him profoundly. Needless to say, the next morning, a train with the precise rakes that needed to be there as per our demand was positioned as promised, much to our delight and much to the dismay of the two Units which were supposed to move prior to us. We ended up reaching Patiala well in time for our Raising Day, with all ranks of the battalion acknowledging my initiative and congratulating me for the miracle!

There is no comparison to the dedication displayed towards any given task in the Army, and this in the real sense is the biggest strength of the Indian Army. I am yet to see any other organization so serious about any assigned task, and I am making this statement in all seriousness since any other organization would prioritize the task based on its importance. However, the Indian Army would give one's 100 per cent even if the task is most inconsequential. The spirit of giving its 100 per cent to all the tasks assigned is amply displayed by the famous military proverb that even if you have to cut the grass, you should be the best in business. Let me give you an example by way of an incident which always comes to my mind whenever I talk about the 'spirit behind any assigned task' in the Indian Army. Years later while I was commanding my battalion, the Corps Commander's son was getting married, and at the last moment, he refused to sit on a horse; instead, simply instructed his father that he would only get married if he could go to his beloved's home on an elephant as part of the ritual. Now there was panic all around, and finally, my battalion was tasked with procuring an elephant overnight; and the task was personally delivered to me by none other than the Brigade Commander himself. Now there was no option but to hold an urgent conference in the middle of the night. Each company was sub-divided into smaller sub-parties and launched into various directions in search of the all-mighty elephant. As the Commanding Officer, I was getting hourly progress reports from the search parties. It was almost 4 a.m., I was to give a report to the Brigade Commander at the first light, and we all were aware that the Brigade Commander never appreciated a 'no' for an answer. I was calling up my Subedar Major[28] time and again, to instruct him to personally motivate the search parties. At last, I heard the beaming voice of the Subedar Major saying that the task had been accomplished, and I was happy to hear that the boys were finally on their way back along with the massive giant. My intent of narrating this incident is simply for the reason that the spirit of the Indian Army, which is appreciated by one and all, is primarily because of this 'never say die' attitude of our soldiers. A task is a task, irrespective of its nature or circumstance or its relative importance. This singular attitude of the Indian Army makes us stand out, and

---

28 Senior-most JCO in the battalion.

even if a boy falls into the well, the Indian Army is the first to reach with a rope. There is no better example of a task-oriented entity than the 'Indian Army,' and this in the true sense is our strength for which the government calls upon the Indian Army at the drop of a hat. From flood relief to the running of trains, or anything under the sun, I can assure you that the Indian Army can do it all, and that, too, better than the original stakeholders. Please try and imbibe this singular trait of the Indian Army, and I can assure you that the sky would also not be the limit! Your dedication in life has got to be like the Indian Army that managed to search for the elephant in time to ensure that the Corps Commander's son was no longer single in life!

9

THE BUS THAT CHANGED MY LIFE

By now, I had put in five years of military service, and these formative years in my career eventually served as the ideal foundation for bigger responsibilities in future assignments. However, in my personal life, there existed a void in terms of a life partner, which my parents were working overtime to fix. Every time I would go home on leave, a long list of marriage proposals with photographs would be waiting for me. While, on an average, an officer in those days would get married generally in their early twenties, I had instead decided to take my own time. I wanted to complete all my mandatory military courses of instruction prior to launching into a sabbatical; there was also the insecurity attached with losing the freedom of bachelorhood deep down. Therefore, on some pretext or another, I had been successful so far in deflecting all such matrimonial proposals. However, my parents had seemingly passed a resolution to ensure that I would no longer be single in life. I was planning to go on a short leave after a gap of almost six months and already dreading the long lecture my mother would give me about how it was high time for me to settle down in life. On every such occasion, I would usually try and take the shelter of my profession. However, this card, too, had outlived its utility because my mother once spoke to my Commanding Officer, and he advised my mother to arrange my marriage at the earliest. With these thoughts in my mind, I boarded a bus from Patiala, which was my duty station, to Hisar for the further journey to my hometown. Normally while proceeding on leave, I would take a bus since there were no direct trains for my hometown from Patiala. Even the bus was not direct, but it was faster since I could manage to reach home in about six hours while the train would take almost eighteen hours. This was the year 1989, a time when nobody was in a hurry, and accordingly, most places did not have direct connectivity either by road or by train. However, life was much happier then; everyone enjoyed the journey, and talking to strangers was a done thing unlike the stiff upper lip of today's generation.

I reached Hisar by evening, and unfortunately, missed the last bus to my hometown. Despite it being only seventy kilometres from Hisar, I had to spend the night at Hisar at my uncle's place. Thus, I am indebted to Haryana Roadways since I met my better half as a result of missing the last bus, which allowed me to meet my wife. She was in her final year at the Haryana Agriculture University, Hisar. I was there that evening and, so I decided to go and meet her in the university hostel as part of a marriage proposal through a common relative. She was still in her teens and looked even younger than her actual age; at first sight, I was convinced that she was too young for marriage! However, what particularly struck a chord with me was her beautiful, charming face with the most innocent smile I had ever seen. I saw her stepping down the stairway and it was like the first rain of the season. Her hair swinging with the wind; Oh! She was truly breathtaking. We exchanged a couple of letters thereafter. I made sure that I missed the bus the next couple of times, and this facilitated a few more meetings. Finally, what brought me down on one knee was her unconditional love for everything on earth. She was like simplicity and innocence personified, with beauty at its best; I was left with no option but to lose my heart on that particular dinner date! Post the unconditional 'yes,' we got married within a few months. Unfortunately, immediately after our wedding day, I was recalled from leave to join my duty station, and that is when I realized why we were always briefed by our seniors to 'remember you are married to the *olive green*'.[29]

The reason for my recall from leave was a change-in-command of our battalion, which was a ceremony wherein the outgoing Commanding Officer would hand over the full strength of personnel of the battalion to the incoming Commanding Officer. Colonel Uppal was taking over the battalion from Colonel Jaswal, and being from another battalion, the new Commanding Officer hardly knew the officers. As a result, it was mandatory for all officers to be present for the 'Change of Baton' ceremony. In every unit of the Indian Army, as is also applicable in any other organization, every time the boss changes, there is a complete shift in functioning based on the style of the new Commanding Officer. Even before the new Commanding Officer had properly settled in, I gathered all my courage and walked into his office, requesting a few days' leave. I am sure everyone can understand my urgency since I was barely married and had hardly spent any time with my wife. Colonel Uppal was surprised with my request; rather than congratulating him on his taking over the Command, I was instead requesting him for leave. Our new Commanding Officer carried himself with a very serious demeanour and was a no-nonsense kind of individual. He instructed me to immediately proceed to Mahajan, a godforsaken place in the middle of nowhere in Rajasthan, to conduct the mortar firing

---

29   A reference to the colour of the uniform of the Indian Army.

of our battalion. You can well imagine the plight of a newly married young officer who walked into his Commanding Officer's office with a request for leave and left instead with instructions to proceed for mortar firing to Mahajan. There was no way that I could have disobeyed my Commanding Officer, so I packed my bag to move to Mahajan, and in the Officer's Mess later, I was the butt of all jokes of the officers the entire evening. However, I had the last laugh, as I made sure that the convoy next morning was moving via Hisar, and there I met my wife in her hostel with a fleet of Army vehicles lined up on their way to Mahajan! The girls in my wife's hostel were so fascinated with the camouflaged vehicles that some of them even came forward to shake hands with my soldiers. This decision of mine perturbed the Commanding Officer since earlier, all military convoys had moved via Bhatinda to Suratgarh, and then onwards to Mahajan. A court of inquiry was ordered to investigate the circumstances under which I changed the convoy route for going to Mahajan. In the entire process, by way of ordering the inquiry, I was made to feel as if I had committed the gravest offence. The officers in the battalion advised me to apologize to the Commanding Officer, but somewhere deep down I was not convinced that I had committed any offence. From my perspective, I was the Convoy Commander, and I thought it was my prerogative to decide the best possible route for the convoy.

In the Army, a court of inquiry is a committee which has to give a fair chance, and thereby record the statements of all witnesses prior to arriving at any conclusions. Like in the eyes of the law, nobody is guilty until one is proven to be guilty. Therefore, everyone gets a fair chance to put across their viewpoint prior to the inquiry, and only after having heard both sides do the committee finally makes its judgment. I had many advisors, and every evening all of us would sit over a drink in the Officers' Mess to decide the best course of action available to me under the circumstances. All such discussions would digress after a couple of drinks, and invariably, we would end up criticizing the new Commanding Officer. By and large, the officers were not happy with the new Commanding Officer's style of functioning. However, they would only have the courage to talk about it in the evenings with their tongues in their glasses. Realizing that such discussions were not heading anywhere, I decided to do my own homework and prepare the statement that I was to present before the inquiry. A young officer with five years of service was facing the first court of inquiry of his life, and yet, my morale was high, and I had decided not to plead guilty in the inquiry. This is a system initiated by the British, in which the accused has the prerogative to plead either 'guilty' or 'not guilty' when summoned by the Inquiry Committee, and all further actions are dependent on the call made by the accused. In case the accused pleads guilty, the inquiry simply recommends the punishment for him; otherwise, the case is open for further investigation.

I was not lucky like in the present day, wherein everyone is in possession of a Global Positioning System (GPS), so I went to the market and purchased a road atlas of Punjab, Haryana and Rajasthan. That evening, rather than going to the Officers' Mess for dinner, I decided to continue my research on the road atlas in my room, as I was to appear before the Inquiry Committee the next morning and make my statement. Major Sukhbir was handpicked as the Presiding Officer of the Inquiry Committee by our Commanding Officer since he was highly qualified in military law. I got a call in my room from our Second-in-Command, who once again advised me to apologize to the Commanding Officer and finish the case there. I assured him that I would definitely consider his recommendation, but in my heart, I had already decided to face the inquiry. In any case, it was too late since the entire battalion was aware of the inquiry, and the only option left for me was to fight my way to justice. Realizing that I had not come for dinner to the mess, all the bachelor officers got my dinner packed and visited me in my room. All of them assured me of their support. I was touched by their gesture. I spent that night deciding my way forward, and by next morning, I was mentally prepared to plead my case in front of the Inquiry Committee. The moment I entered the makeshift office of the Inquiry Committee, I was asked to take my seat and the convening order was read out to me. Convening order is the authority for the Inquiry

Committee, which specifies in exact terms as to what the inquiry investigates. Hearing the allegation mentioned against me in the convening order, I immediately declined and pleaded 'not guilty' of the same. I was asked to make my statement on the entire issue, and I simply started with a one-line statement, 'If savings to the state exchequer are an offence, I am guilty.'

Major Sukhbir could not understand what I meant and asked me to explain in detail what I implied by the words, 'savings to the State Exchequer'. I placed on his table a calculation sheet which I had worked out most meticulously the previous night, giving out a comparative study of the detailed expenditure incurred by virtue of our convoy moving on both the routes. I requested the Committee to attach my calculation sheet with their report and declined to make any other statement. Major Sukhbir asked me to wait outside and walked up to our Commanding Officer with the detailed calculation sheet presented by me. Incidentally, nobody had worked out the total distance to Mahajan on both the routes, and by virtue of my decision to move the convoy via Hisar, it resulted in savings to the exchequer as the route via Hisar was shorter by thirty-five kilometres. I could read the tense expression on Major Sukhbir's face once he returned, and he quickly concluded the inquiry and ordered me to carry on. The entire exercise of me presenting this incident to my readers is to drive home the point that if you are convinced of your decision, and yet it is being questioned by the environment around you, it is extremely essential to 'hit the nail on its head' without mincing your words or actions. However, it is more important to identify the correct nail prior to hitting it on its head. I, as a young officer, understood the wisdom of common sense. The road atlas provided me with the knowledge; however, knowing how to use the knowledge derived from the road atlas was my wisdom. This was the most important learning in the entire episode, and thereafter, I was the champion of utilizing the resource called 'knowledge' to its correct usage, thereby applying my wisdom. It takes an entire life to understand this singular truth—that knowledge is pointless in case you do not know how and when to utilize it. We can understand the difference between knowledge and wisdom from that simple phrase everyone says, 'He is a wise man,' whereas nobody says that he is a knowledgeable person. Therefore, understand the fact that knowledge without wisdom is pointless. Understand the difference between knowledge and wisdom. Be the wise man rather than being full of knowledge and not knowing what to do with that knowledge. Towards this end, *hasten slowly, love unconditionally and think logically.*

10

GOOD OLD SOLDIER SIMAR SINGH

I was commanding Number Three Platoon of 'Alpha' Company', and as luck would have it, somehow, all the defaulters of our company had managed to assemble in my platoon. Our Company Commander, looking at the unique 'dirty dozen'-type soldiers of my platoon, would in a lighter way refer to me as the 'leader of the defaulters'. However, merely referring to my boys as a bunch of defaulters would not do justice or complete the picture, since most of us were champions in our own field. No battalion team was complete without representation from my platoon; every soldier in my platoon had some unique quality and was an expert in his own domain. My boys were physically very fit and stood out merely for their physical robustness, when compared to the other soldiers in our company. There was never a dull moment for me while commanding the elite third platoon of 'Alpha' company.

The Company Sergeant Major was always complaining to me regarding the conduct of my boys, and on every such occasion, I would act like an umbrella, trying to shelter my children. The young officers of other companies would always wonder how I was able to manage this bunch of undisciplined soldiers of my platoon. I believed in my boys and tried to explain to everyone that every soldier of my platoon was unique and gifted, and that I was confident that each one of them would perform in an exemplary manner during actual operations. They may be having issues specific to conventionally accepted discipline standards, but professionally, they were most competent, and I was very happy to lead them into war whenever required. I made it a routine to dine with them at least once every week. On every such occasion, after a couple of drinks, their real talents would start showing. Till today, I cherish those wonderful evenings spent with my boys in the platoon.

It was an exceptionally close-knit bunch of thirty odd soldiers whose morale used to be sky high and they were ever ready to take on any task assigned to them. The main issue which plagues the morale of the Indian Army is that of leaves since every soldier looks forward to timely leaves, and when that can be assured, nothing else matters in life. The leave's connection with the morale of the soldiers was well understood by me early in my career, and therefore, I had always ensured timely leave for every soldier of my Sub-Unit, irrespective of his Unit commitments. In the case of the third platoon, I had allocated dedicated vacancies to each section[30] of my platoon, and as a result, the boys decided their own leave by rotation. In most Units, the leave plan is prepared by the headquarters, and is thereafter disseminated down to the boys, which never works; on the contrary, I always believed in the system of 'leave vacancies' being allocated to every Sub-Unit and allowing the boys to decide their own leave plan. Over the months, our association grew manifold as we were practically training together, eating together, and playing together, and I was a thoroughly satisfied Platoon Commander. Havildar Simar Singh was the driver of my Infantry Combat Vehicle (BMP).[31] Unlike a normal four-wheeler, driving a BMP is considered to be the most challenging job as a driver, and Simar was one of the best in this business. The biggest challenge for a BMP driver used to be crossing the Bridge Laying Tank at night over a water obstacle. The bridge was used for crossing any natural or artificial obstacle created by an enemy during operations. The crossing of the bridge

---

30   A group of ten soldiers.

31   BMP is a Soviet amphibious tracked infantry fighting vehicle.

at night used to be a herculean task since the bridge was only as wide as the BMP itself, and a slight mistake in the alignment of the BMP would result in it falling off the bridge. Therefore, this used to be the litmus test with respect to the professional capability of any BMP driver, thereby at times causing injuries to the soldiers inside the BMP. During daytime, the crossing was relatively easy. However, prior to crossing at night, I have often seen the best of the drivers heading for our unit temple to seek the blessings of the Almighty.

In our training area, we had constructed a cemented bridge with the exact dimensions of the operational equipment. The drivers would often go wrong on the bridge, and therefore, to save the operational bridge from damage, this cemented structure was constructed for practice. Initially, a driver used to be trained to cross this bridge during daytime, and thereafter, gradually, he would be allowed to start night crossings. Even the most trained drivers would often go wrong. The annual inspection for every unit was an exercise carried out to assess the fitness of the unit for war, and its relevance was of paramount importance. Our battalion's annual inspection was fast approaching, and everyone was working hard to be at their best. We were told to put additional efforts and energy into our respective tasks since our Brigade Commander was known for shooting surprises at the spur of the moment in every such inspection. Therefore, a month prior to the inspection, leaves of all ranks were stopped and the battalion started preparing on a war footing for the event. The Brigade Commander arrived on the designated day for the inspection and the first order he gave immediately on arrival was to mobilize our battalion for war. Mobilization is a standard procedure, wherein every unit is tasked to mobilize within six hours, which includes loading of ammunition and wartime stores as well as dumping of rear stores in specified barracks under lock and key to ensure their safety. All our BMPs were to move into the harbour area, which was next to our training area. The harbour area was designated for safety against enemy air attack, and the complete array of operational equipment used to be well spread out and under camouflage in the vast area specified as the harbour. Once the Brigade Commander was given the completion report of our mobilization, he walked into the harbour area and tasked all our BMP drivers to demonstrate the crossing of the Bridge Laying Tank at night. Since it was already close to sundown, we mustered all our drivers with their BMPs and marker lights were fitted on the cemented bridge. As a result of the focused training in our battalion, fifty-two drivers, as per the authorization of a battalion, crossed without any mishap, which was commendable, and we all were quite relieved.

Just as we thought the Brigade Commander would call off the exercise, he summoned all our drivers and asked for a volunteer to cross the bridge, blindfolded. Now this order was beyond our wildest nightmares and was bordering craziness. To my utter surprise, the hand that went up was that of Simar Singh, my driver! There was nothing left for me to do but look towards the sky, hoping and praying for a miracle. Initially, we thought that the Brigade Commander was probably only trying to check our level of confidence. However, pretty soon we realized that he was extremely serious, as Simar was blindfolded. The moment Simar's BMP started moving forward towards the obstacle, I remember the entire battalion standing in pin-drop silence, in anticipation. I understood that the entire battalion was silently praying as hard as humanly possible. As the BMP approached the obstacle, I was holding my breath, and I presume, so was everyone else. We were at a distance, and therefore, could only hear the distinct sound of the BMP track hitting the cemented structure with some dust flying, and in a matter of seconds, the BMP was across the obstacle. The loud war cry by the battalion at that moment cannot be forgotten and thinking about this incident still gives me goosebumps! Naturally, the Brigade Commander was extremely pleased with this spectacular feat, and he gave Simar Rs 5,000 on the spot while complimenting our Battalion Commander for a very high standard of operational readiness. I, personally, was aware of Simar's professional capabilities but never knew that he could do the feat blindfolded! The moment he came up to me, I could not utter a single word and simply hugged him. I could see the emotionally charged Simar with his eyes moist.

Once we reached our company, Simar told me that when the Brigade Commander threw this open challenge, if nobody from the battalion would have volunteered, it would have been an insult to our paltan.[32] The very same day, our Battalion Commander announced Simar's promotion to the rank of Naib Subedar,[33] which otherwise may not have been possible, owing to the image of an 'undisciplined soldier' attached to his name. The next morning, I deliberately approached my Company Commander with a big smile and saluted him with the loud words, 'Leader of defaulters reporting, Sir,' and was extremely happy with his response, 'Yes, I believe, this country needs your defaulters!'

In the Indian Army, we judge a soldier by his expertise specific to his trade. By trade, we mean the job the soldier would be performing in war. For example, in case he is going to drive a vehicle during operations, his trade ought to be a driver, and in case he is going to fire a gun, his trade has got to be a gunner. The training as a routine is also focused on his trade, and over the years, a soldier becomes the master of his trade like Simar Singh. The word 'trade' in the civil sense denotes your profession, but taking a clue from the military, I request you to be the best in your particular expertise. If you are an expert in your profession, you will be successful in life; in turn, this success will translate into happiness, and I can assure you

---

32   Battalion.
33   JCO.

that you will be contented in life. Many youngsters share with me that they want to be happy in life, but they say so without knowing the real meaning of happiness. Nobody can be happy without work, and if you want to work, you have got to be the best in your business. This, and only this, can satisfy you in life.

11

BREAKING THE THREAD WITH HIS CHEST

In a lifetime in the Indian Army, one comes across situations demanding ultimate dedication and decisiveness not only in yourself, but also, from the men you command and lead. And within this gamut, one can choose to wait around and hope for good things to happen or leap forth and put in the work and effort to achieve that dream. While opportunities and options can always be carved out from the space around, we cannot always control the outcome of our efforts. It is only when one can create that illusion of control and optimize the given opportunities and options that the symphony of achievement is created. The soldiers of the Indian Army, who form the bedrock of the organization, are drawn from the vast and vivid diversity of our country and come with the concurrent multiplicity of cultural and ethnic backgrounds. And yet, the moment they get inducted into the life of uniform in the Unit or Regiment, all this diversity coalesces into a simple mantra of motivation in the three golden words, naam,[34] namak,[35] and nishaan.[36] It is these three defining symbols that makes a soldier prepared to give his best and even his life, be it in a battlefield or in a playing field, contesting at both places for the izzat[37] of his company, his Unit, his regiment and ultimately, his country.

With the passage of time and with my growth in seniority in my battalion, I was ultimately given the onerous responsibility of commanding a company. Being a Company Commander is probably the most important event in the life of any officer since it is technically the first command opportunity with the mandate to fight a battle independently, if required. As a Company Commander, one gets the occasion to directly train, motivate and lead a group of soldiers, both in war and peacetime commitments. Also, a company as a Sub-Unit organization develops a

---

34 The regimental name.
35 The regimental loyalty.
36 The regimental flag.
37 The regimental pride.

character and ethos of its own, which is manifested in its troops and officers. The regimental izzat is developed in the nursery of its companies. Consequently, a healthy inter-company rivalry co-exists within the more global regimental bonhomie of the Unit. And it is this inter-company rivalry that goes a long way in developing and shaping the ultimate regimental and national pride in the soldiers. I was handed the command of the Alpha Company of my battalion, which was my parent company since I had been commissioned into it as a Second Lieutenant. Consequently, it was more of a homecoming for me, and I felt I was on my home turf. I knew almost all the men in the company. I had seen many of the senior NCOs[38] and JCOs[39] grow along with me in service. In fact, I even knew the familial backgrounds of most of the men in my company, having spent quite a few years with them in both field and peace deployments. We shared a mutual trust and empathy, which was one of the cornerstones of our relationship. Being deployed in a peace profile, our Unit was constantly under various tactical and operational trainings. One of the most important facets of such training is the various sports events that are organized to keep the fitness levels and motivational drive of the troops at an optimum level. It is in these sports events that the

---

38 Non-Commissioned Officers (Sergeants/Havildars, Corporals/Naiks).

39 Junior Commissioned Officers (Subedars, Naib Subedars).

inter-company rivalries peak to almost feverish pitches, at time even leading to interventions from senior officers of the battalion to calm down situations and tempers.

My Alpha Company, for some reason, had not been able to secure the Unit Championship Banner for quite a few years now. One of the first objectives that I had penned down for myself on assumption of the command of this company was to work towards and win the Championship banner. This required a huge amount of investment in training and performance in the various tactical and sporting events that were scheduled in that training year.[40] But I was confident that if I could somehow get my men motivated enough, the objective could be achieved easily. With this as the primary aim, I convened the Company Sainik Sammelan[41] and declared it as my intention to win the Unit Championship Banner that year at any cost. My company voicing my intention in unison, led by my senior NCOs and JCOs, decided to work towards this objective in a diligent manner. A detailed and meticulous training schedule was prepared, and we trained and participated in all the events of that year by using a very methodical approach. Selected performers were spared duties and fatigues and were provided additional dietary supplements. The senior NCOs and JCOs of the company took on the responsibility of keeping the morale of the company at its highest for

---

40 1 July to 30 June.
41 Open House Forum.

each event. In fact, each soldier eagerly contributed his bit so that the performance of the participants could be the best. Our planning and preparations didn't go unnoticed, and other companies, too, were bitten by the competitive bug. The incumbent Champion Company was the Charlie Company under the command of Major Kailash, and he declared to his company that under no circumstances would he or Charlie Company allow Alpha Company to take away the Championship Banner. As event after event progressed with varying results, the inter-company rivalry started getting more and more intense, and in fact, came to such a point that soldiers of Alpha and Charlie companies even stopped talking to each other for some time. However, towards the end of the training year, it was realized that both Alpha and Charlie Company were more or less tied in points for the honour of the Unit Championship Banner; the last event, the cross-country run, would be the tiebreaker. Now, the cross-country run was an extremely important event which churned out the maximum points, and the winner of this event could contribute a huge number of points to the kitty of the company. Hence, the winner of that year's cross-country run would undoubtedly secure the Championship Banner.

The cross-country run is one of the most primal, and yet, demanding sporting event, which tests not only the physical stamina of the participants, but also to a huge extent, the mental endurance and spiritual resilience of the runners. It is a sport where one competes not just against one's competitors, but also against one's own self. A time comes during the run when the body and mind keep asking you to give up and stop, and yet, it's your spirit which keeps driving you forward one step at a time till you reach the fabled tape or thread at the end of the route. And the entire time while one is running, myriad thoughts go through their mind, some negative and some positive, some demotivating and some inspiring. It is this play of thoughts and the synergy with the body that determines one's success in a cross-country run. Unlike other sports requiring various gadgets or equipment and with a number of rules to adhere to, the cross-country run requires just the runner, the road, and the sky above. As the cross-country run was drawing near, there was a palpable tension in the company on the possible outcomes of the event. We all knew that only a stellar performance as a company and an outright win in the run would make it possible for Alpha Company to win the Banner. So, it was imperative that a probable winner would have to be identified and trained to lead the charge of the company and achieve the objective. It was during these confabulations in the company that Havildar Rajender declared that come what may, he would 'break the thread' that year. As is customary, during a cross-country run, a thread or tape is strung across the finishing line. The winner needs to run through this thread or tape to win the event. This running of the winner through the finishing thread or tape is what is referred to as 'breaking the thread with one's chest'.

Though Havildar Rajender had proclaimed and taken it upon himself to do the job, we in the company were a little skeptical of the whole issue. The primary reason for it was that Havildar Rajender was a wrestler and had the body built for that sport. He was on the heavier side with a lot of muscular features, something that is generally not considered to be an advantage for a long-distance runner. My senior NCOs and JCOs confided in me this dilemma, and for a few moments, I, too, had my doubts. But I had known Havildar Rajender for a long time and was certain of one trait in him. He was an extremely focused individual, and given a task, he would put his complete mind and soul into it. So, in spite of my apparent reservations, I decided to give him the opportunity to try and contribute to the company's fortunes. Two things drove me to this decision—one, the sheer audacity of determination in Havildar Rajender in taking up such a challenge; and two, if he was somehow able to achieve what he had set himself up for, it would do a tremendous amount of good to the larger professional and military ethos of the company and Unit as a whole. So, I instructed the Company Havildar Major[42] and the senior JCO to extend all possible help and assistance to Havildar Rajender in accomplishing his objective.

The company set itself up for the oncoming event. Havildar Rajender started on an extremely severe and

---

42 Same as Company Sergeant Major.

exacting training regime. He would go on practice runs at almost all times of the day. He would run with sacks filled with sand on his shoulders and with weights tied to his feet to increase his endurance capacities. We all contributed by chalking out his diet plan and serving him a proper diet. The youngsters of the company would run along with him in stretches, timing him and continuously motivating him. Soon, I saw a perceptible loss of weight in him, and gradually he developed the lanky shape of a long-distance runner. His stamina kept improving with each day as did his timings. While we all as a company were with him from the sidelines, it was amazing to see the incredible punishment to which Havildar Rajender subjected his body to achieve the right ability to accomplish his dream. However, just a week before the event was to take place, we received a telegram from Havildar Rajender's home, informing that his father was extremely sick and admitted to a hospital and that he was required to reach home at the earliest. This was a big blow to all of us. My senior JCO came to me to seek guidance on whether we should inform him or keep silent till the event is over. I decided that it would not be morally correct to keep such an important news hidden from him, and thus, I called him into my office. The moment I read out the telegram to him and told him that all arrangements had been made for his departure by train that afternoon, I saw a small tear in the corner of his eyes, signifying how much he loved and missed his father at that moment. But in just a couple of seconds, he composed himself and told me, 'Saab,[43] main abhi nahi jaa sakta. Mujhe ek hafte mein unit ki cross country championship mein bhagna hai. Aap mujhe 15 din chutti de dena event ke baad' ['Sir, I can't go now. I have to run the Battalion Cross Country Championship in a week. Kindly grant me fifteen days leave after the event.'] His reply stunned me. I immediately replied, 'Par tum iss mushkil waqt mein apne pita ke pass nahi hona chahoge?' ['But wouldn't you like to be next to your father at the time of this medical emergency?'] His next statement was even more inspiring. He said, 'Saab, agar aaj hum ladai mein hote aur dushman darwaze par khada hota, toh kya aap mujhe apne saathiyon ko chodh kar jaane ko bolte?' ['Sir, if we had been fighting a battle now and the enemy were at the gates, would you still have asked me to abandon my post?'] I felt really humbled by the devotion and loyalty of this finest example of an Indian soldier. Though we were not in any operational or battle situation, yet just the naam, namak and nishaan of the company in a sporting event had the same relevance for this man. It is this indomitable spirit of the soldiers of the great Indian Army that makes it one of the finest fighting forces in the world today. Havildar Rajender got back to his training schedule and we in the company decided not to overtly burden him with his problems any further. However, I immediately sent a senior NCO of the company to his hometown the

---

43 A form of address by soldiers to officers.

same day with the instructions to be by his family's side and render all necessary assistance in the treatment of his father.

The night before the cross-country run, we were all wreaked with anticipation and anxiety. It is said that to have a good run, one must have a long, relaxed sleep the night before. But, unfortunately even after trying hard, none of us could manage to get a wink of sleep. Feeling anxious, I went down to the company lines to check up on the men and found almost everybody awake and tensed. I walked down to Havildar Rajender's cot and found him awake, too. He got up to salute me the moment I neared him. I waved him off and asked him to keep lying down. Taking a camp stool, I sat down at the head of his bed and informed him that the NCO who had gone to look after his father had called up and informed that his father had responded to the treatment and was well on his way to recovery. This news brought out a fresh tear in his eyes and he thanked me profusely for going the extra mile. He told me how his father, an extremely poor man, had gone through enormous hardships to educate him and finally he got enrolled in the Army. I was really impressed to hear about the humble background from which had emerged this man with such conviction and courage, not often found in much more educated and privileged people. Asking him to try to sleep, I came back to my quarters to try and find some sleep for myself, too. The day of the cross-country race was adorned with a beautiful clear sky with a slight nip in the air. We all

Academy under Officer Rajpal Punia meeting the Soviet Defence Minister at the Indian Military Academy, Dehradun (17 June 1984).

Beating the blues at Tangdhar (15 December 1998).

Major Rajpal Punia along with United Nations observers of various countries at Kenema, Sierra Leone (January 2000).

Appreciation letter by the Force Commander, Major General V.K. Jetley, as received by Major Rajpal Punia after the successful execution of Operation Khukri (18 July 2000).

Major Rajpal Punia along with his soldiers at the Investiture Ceremony at Chandimandir (16 December 2002).

Havan [religious ritual] performed by Colonel Rajpal Punia, YSM, on assuming command of his battalion (April 2005).

Puja [prayer] performed right before the first flotation at the Sirhind Canal (September 2005).

Colonel Rajpal Punia swimming across the canal to motivate his soldiers (September 2005).

Colonel Rajpal Punia being congratulated by Captain Amrinder Singh, the then Chief Minister of Punjab, after the successful completion of the construction of a cenotaph in Patiala. (December 2005).

Colonel Rajpal Punia with his pillars of strength in Patiala (December 2005).

Brigadier Rajpal Punia and his wife, Mrs Anita Punia, offering prayers before taking over the Fantastic Fifth Brigade at Along [Aalo], Arunachal Pradesh (February 2012).

Brigadier Rajpal Punia visiting one of his Units at a forward location in Arunachal Pradesh (March 2012).

The slogan that every soldier of the Indian Army swears by, written on a gate at Taksing (March 2012).

Brigadier Rajpal Punia walking along Subansiri River on his way to Taksing (October 2012).

Brigadier Rajpal Punia crossing the suspension bridge on his way to a forward post on the Line of Actual Control (October 2012).

Brigadier Rajpal Punia standing at the final Indian Post overlooking China. (November 2012).

A volleyball match with the soldiers at the Taksing Unit (September 2012).

A basketball match with the officers at Menchuka (September 2012).

Brigadier Rajpal Punia at the scout camp in Bangkok, Thailand (November 2014).

Major General Rajpal Punia was invited as the Chief Guest at Rajgarh, Churu for the felicitation ceremony of the next of kins of the martyrs (January 2017).

Major General Rajpal Punia leading the Republic Day Parade from India Gate to Red Fort (26 January 2018).

Major General Rajpal Punia YSM saluting the Supreme Commander of the Armed Forces at Rajpath (26 January 2019).

Camaraderie between the Army and the paramilitary force at Sirsa exhibited in this picture while General Punia's division was tasked to vacate the dera of Ram Rahim at Sirsa (26 August 2017).

gathered at the starting line, waiting for the race to start. The runners were all lined up in their company-coloured vests. Alpha Company was sporting their traditional red vests, while Charlie Company, our main rivals, were in their blue vests. Each runner was going through his own personal motions of warming up or psyching themselves up. I tried to get a look of Havildar Rajender before the race started, but he was squatting down with his head between his knees, as if in a deep private prayer to his God. He never once looked up or said a word.

The race started with the traditional blank rifle shot being fired in the air. The runners scrambled down the road towards the various checkpoints which they had to cross enroute. Radio operators, placed at various spots on the route, kept giving us a running commentary on the progress of the race. In the initial two checkpoints, it was reported that Havildar Rajender was leading the pack. The news was met with huge jubilation from the Alpha Company tents. However, by the third and fourth checkpoints, the runner from Charlie Company had overtaken Havildar Rajender, and he was falling behind. I was crushed for some time. I realized that Havildar Rajender must have given his best, but the Charlie Company runner was, after all, the best runner in the battalion, and hence must have overtaken at some point. Somehow, the radio set at the last checkpoint didn't work and we couldn't get an update from there. Now the runners were in the last stretch of three kilometres, with the last kilometre being a straight back-breaking climb to the finish tape. One could see the runners appearing around the last bend and starting the slow climb towards the end. In front, I could see a blue vest leading, followed by a red vest a few steps behind. Gradually, as they emerged into focus, I could see Havildar Rajender following the Charlie Company runner and slowly but steadily closing in on him. With just about fifty metres to go, Havildar Rajender was about two steps behind the leader, and I could see that he was almost at the end of his stamina. White froth was coming out of the corner of his mouth and he was propelled forward simply by his mental strength.

At the last fifteen to twenty metres mark, Havildar Rajender suddenly looked up. I was standing a little ahead of the finishing tape and had intended to run the last few steps with him. He looked at me, gave a wry smile, and then, like a panther, made a mad leap and dash for the finish line, overtaking the Charlie Company runner by maybe just a second. But he did 'break the thread with his chest'. I had reached him by now and he simply collapsed on me. He looked up and said, 'Saab, umeed karta hoon ki aapne jo chaha tha, voh main poora kar paya?' ['Sir, hope I have been able to do what you wanted?'] and passed out unconscious. My hands were shaking, and I had almost gone numb in my mind, seeing this performance. But I immediately came out of my reverie and called out for the medic. We rushed him to the unit medical room and administered some first aid. However, he remained unconscious, and his pulse was falling rapidly. We mustered an ambulance and rushed him to the nearby Military Hospital, where he was immediately taken into intensive care. Luckily, he was diagnosed with extreme exhaustion and dehydration, which was treated efficiently. I waited in the hospital for the next few hours till such time the doctors declared him out of danger and fit to be taken back to the Unit. I brought him along with me in my vehicle. Our entire company was standing at the regimental post (RP)[44] Gate, including some of the other company boys, including the Charlie Company runner and Company Commander. The men of the company carried Havildar Rajender on their shoulders from the gate to the company lines, and Alpha Company duly received the Unit Championship Banner that year from the Commanding Officer, majorly due to the unbelievable run by Havildar Rajender. Once the labyrinth is over, we won't remember how we made it through or how we managed to survive. We won't even be sure whether the tough time is really over or not. But one thing is certain, when we come out of it, we won't remain the same person who had stepped in. Till date, I have never met anyone having as much devotion towards the organization and towards his duty, as was displayed by Havildar Rajender on that particular day. From that day onwards, I have always kept him as a benchmark and a role model to try and emulate his sense of devotion, pride

---

44 RP Gate: the main gate of the Unit.

and all-out efforts. Currently, Rajender is retired but I have still remained in contact with him. He is now employed as a Security Officer in Tata Motors at Jamshedpur, and yet, I have remained in contact with him. Every time I speak to him, I never forget to pull his leg by asking him if he is still willing to run the cross-country. His reply every time would be, 'Saab, Dil se abhi bhi jawan hoon aur agar koi jeet sakta hai toh voh hai Rajender!' ['I am still young at heart and if anyone can break the thread, it would be only Rajender!']

12

SILVER JUBILEE OF A GRENADE SPLINTER

Our battalion had moved to Babina in the year 1993. Babina is a small town in the state of Uttar Pradesh, and I doubt if any of my readers have heard its name. The British had established this cantonment in central India and given it a new name, the 'British Army Base in North Asia', abbreviated to BABINA. That is the origin of this township near Jhansi. Over the years, this cantonment grew into settlements all around, and currently, it is a tehsil[45] of Jhansi district. It is barely twenty kilometres from Jhansi, but the most interesting fact about its location is that while travelling by road from Jhansi to Babina, one has to cross the states of Madhya Pradesh and Uttar Pradesh at least four times. A firing range is located at Babina, which is utilized for heavy equipment firing such as tanks and artillery guns. Our battalion was still in the process of settling down in the new military station with Colonel Chand at the helm of affairs, and we all found Babina to be a very unusual place since it was an imperial British establishment that was turned into a village. Babina was a colonial cantonment around which the village mushroomed.

Ever since we had arrived at Babina, I had developed a severe backache, and had tried all possible medication as well as physiotherapy treatment, but to no respite. Being a sportsperson, it used to be difficult for me to sit outside the ground and see my team losing the inter-unit tournaments. However, the pain was so severe that it used to be challenging for me to even stand. Finally, on the recommendation of my Commanding Officer, I went through all possible tests at Jhansi and was diagnosed with Prolapsed Inter-Vertebral Disc (PIVD), and the surgical specialist advised me to go for an urgent surgery at the Command Military Hospital, Lucknow. I was on the verge of getting transferred to Lucknow when Major Menon of the First Guards walked

---

45 Sub-division.

into the officers' ward of the military hospital as my guardian angel. The first thing he showed me was a 'waist belt' he was wearing for the same ailment. Major Menon was also a good basketball player and we had played against each other in an inter-battalion basketball tournament. He had specially come all the way to get me discharged from the military hospital after he heard about my back problem and likely transfer to Lucknow. Moreover, he introduced me to some 'yogic postures', which, over a period of time, would recover and heal my backache without surgery. I followed his instructions and initially did yoga under his watchful eyes, until I was confident of continuing on my own. Within a month or so, I could feel the relief. Over a period of time, I recovered completely. My reason for writing about it is primarily to request everyone to avoid surgery for PIVD, and instead, try yoga and physiotherapy as much as possible. PIVD is a common physical ailment, wherein the disc in your backbone may shift as a result of a normal or physical activity, after which it starts pressing the nerves passing through the backbone, resulting in unbearable pain. Surgery should be the final option that should be considered after having tried all other possible remedies. Books on yoga mention that specific yogic postures can cure what is commonly referred to as 'slipped disc' when they are practiced over a period of time, provided you perform them under the supervision of an expert. I hope to share a ray of hope and positivity with all the patients of slipped disc. Over the years, I have been able to help out a number of patients suffering from slip disc, and I once again reiterate the importance of avoiding surgery to the extent possible.

While suffering with my lower backache, I had appeared for the Defence Services Staff College examination, which is quite a tough competitive exam. In my case, the difficulty was more, as I had appeared without much preparation. God was kind, however, and I qualified. And so, it was a time for celebrations for the entire family since the qualification meant that besides good promotion prospects, I would enjoy a year of quality life at Wellington, which is next to Ooty in Tamil Nadu. Just prior to my move to Wellington, the annual field firing of the battalion approached, and despite being on leave, I was recalled to conduct the grenade firing since I was an instructor in small arms and had also conducted grenade firing many times in my career. Unfortunately, the soldiers participating in the grenade firing training were going to throw grenades for the first time, since most of them were young soldiers fresh out of the centre. I had never seen soldiers so petrified while throwing a grenade. I could virtually see them shivering the moment they held a grenade in their hands.

My senior sergeant was amazed at this and mentioned to me that it would be difficult to train soldiers who were so scared. Looking at the hesitation displayed by those greenhorns, I explained the mechanism of grenade explosion to them, and highlighted the fact that a grenade is like a stone till you remove its safety pin, and even after removing the safety pin there is ample time to throw it, since it takes five seconds for the grenade to blast once it is launched. However, there was no improvement, and the sight of the scared soldiers was becoming personally disgusting to me. Finally, to send a message across, I did something reckless, deciding to sit outside the bunker while they were lobbing the grenades. My aim was to send a message that it was pointless to be scared, looking at how their Company Commander was seated in the open. Though I was motivating my soldiers with my personal example, in the process, I was actually violating a basic standard operating procedure of grenade firing.

The grenade lobbing continued, and the splinters were flying over my head. I was told that everyone had finished and very few grenades were left. So, I asked the instructor to prepare a balance of the grenades left and give a second chance to any volunteers. When the last detail had thrown a grenade, I suddenly felt a hammer-like force hitting my knee. For a moment, I felt extreme heat and excruciating pain, followed by complete numbness in my right knee. I tried getting up despite the extreme pain, and that was when I noticed a small hole in my trousers. Despite severe pain, I tried to put on a brave face since I noticed that my company was watching me. The nursing assistant saw the wound and applied dressing. I felt a little better. By then, breakfast was ready, and since everyone had finished their grenade-throwing, I asked my company to have breakfast and joined them for the same. The information of my injury reached our battalion headquarters and I was ordered by our Commanding Officer to proceed to the military hospital for medical examination by the doctor. On reaching the military hospital, an urgent X-ray test was carried out, and I was shocked to see the image with two clear splinters in my knee joint. The Surgical Specialist, Major Gupta, smiled and asked me if I had an empty stomach. I smiled as I told him, that I had had my breakfast post the injury. Major Gupta was horrified as my injury demanded an urgent surgery and I had to have an empty stomach for the same. The rotis[46] I had for breakfast with my troops turned out to be the most painful experience since now the surgery had to be undertaken by administering local anesthesia. Even today, when I recollect the horrible experience of my knee surgery under local anaesthesia, my entire body shivers. The doctor was a gentleman; however, he had no option but to dig down into my knee because unfortunately the size of the splinter was too small to catch his eye. He tried all his steel gadgets, including the hammer, but unfortunately could not home in on the target. Finally, the poor patient lying in severe pain suggested the infantry technique of homing on to the target, which was kind of celebrated by the entire medical team present in the operation theatre. Despite unbearable pain, my mind was at its best, and I simply asked the

---

46   Indian bread.

doctor to insert his steel into my knee and try his X-ray to verify whether it was hitting the target. Finally, the X-ray confirmed that the steel rod was bang on, and one more similar pincer was inserted to design a pair of tongs to pull out the splinter! Even after twenty-five years, I get goosebumps writing about it, and one can imagine the pain I would have gone through when Major Gupta pulled out those inhuman surgical tongs. I fainted despite being under local anesthesia, and when I regained my senses, Major Gupta personally complimented me from the core of his heart for being his bravest patient. But he also shared the unfortunate news that the second splinter was too deep to be pulled out. So, a trophy of that 1996 field firing is still preserved in my knee, and each time I stand, a little cracking sound reminds me of the foolish act of a young Major seated on a camp stool (field chair) in the open with splinters flying all over. The only silver lining was that now a soldier was complete. It is a common saying in the Army that in order to be a battle-hardened soldier, one has to have a bullet injury. In my case, the grenade splinter was testimony of my adventurous life in the Army.

Finally, I wish to write of the most painful experience of surgery that there is a very thin line between 'being brave' and 'being foolish'. Today, when I look back, I feel that it was childish of me to have been sitting out during the grenade firing, thereby violating the laid-down standard procedure of the Indian Army. God punished me for this act, and even eating breakfast post the injury was part of

God's act, without which I think the punishment would not have been complete. Let me try to put across the entire incident more scientifically. I am sure you all must have read Newton's laws of motion in school. What always fascinated me was Newton's third law of motion, which exemplifies the entire philosophy of life in the simplest way. Newton's third law states, 'Every action has an equal and opposite reaction.' This also happens to be a fact of life; every time, there is an act of God that will occur to balance your action. If you do good, I can assure you that good will happen to you. And if you do something wrong, it will also come back to you in a similar manner so that you regret your original act for your entire life. Therefore, always remember Newton's third law of motion, or can I say 'law of life', and I can assure you that you will be the happiest person on this earth. Just like in the Army, where we need to obtain a 'no dues' clearance while proceeding for posting, you will also need the same clearance prior to bidding goodbye to your near and dear ones. Nobody has ever witnessed the next life, so all your accounts will have to be cleared in this very life. So, for your departure journey to be smooth sailing, your journey of life should be positive and virtuous.

13

BOXING AT THE RIGA

Following my grenade firing mishap, I was sent on four weeks' sick leave post my surgery, and was most likely to be downgraded medically, which in essence would have ensured goodbye to my chances of proceeding to Wellington for a year. I returned to my hometown during my period of sick leave and continued with all sorts of treatments to ensure that I recover so that the family sojourn to Wellington was not compromised. That is when the local knowledge of an Ayurvedic[47] herb

---

47   A traditional Hindu system of medicine.

helped me immensely. I would like to share this secret with you all since this, at least to my mind, is the best antibiotic on earth. The root of the Indian berry plant, once boiled in water, turns the water red, and this elixir can be used to wash any injury. I can assure you of a complete recovery in no time. One can also simply drink it like tea to cure any type of internal ailments in the human body. My family has been practising this since my grenade injury, and the results can only be seen to be believed. I recovered almost completely within four weeks, with only a scar on my knee visible, and upon seeing the progress, Major Gupta signed my warrant for Wellington. In the Army, a warrant is like a document for railway travel, and once you deposit it with your railway booking, they in turn provide you with the necessary railway ticket for your destination. Seeing the warrant in my hand, there was a celebration in the family. Don't look so shocked; this isn't the warrant that gets you behind bars but one that helped me climb another step in the stairway to success. We finally decided to board the Wellington Special from New Delhi to Coimbatore in Tamil Nadu. The Wellington Special was a specially designed train, and it was run only for officers selected to attend the Defence Services Staff College at Wellington.

The train would accommodate all officers and their families, and this journey on the Wellington Special was one of my most memorable experiences that I will cherish for the rest of my life. Firstly, the spirits of everyone on the train were sky high since most of the officers were meeting after their Academy days. In fact, the satisfaction of success was visible on our faces since each one of us had worked hard to clear the merit-based written examination for the Wellington course. We realized that we were going to be a big family for the next one year, and so, while on the train, the officers were busy catching up on the past decade, while the ladies were busy making friends. The children were the happiest lot, playing all through the journey as the architect of the railway compartment designed especially for the 'Wellington Special' had provided a huge empty area in the centre of the bogie, while all the berths were to the sides. This had created a huge playground for the children in the heart of every compartment, which ensured that they did not trouble their parents at all. One could not have asked for a better journey, with the drinks bar functional in the evening, along with mouthwatering meals cooked by the best chefs available on the train. I was meeting my best friend, Paritosh Pant, for the first time since we had graduated from the Academy in 1984, and twelve years was a very long period to catch up on. Luckily for us, our wives also struck a chord on the very first meeting. Our children were almost the same age, and as a result, Gotu, Pranav, Arjun and Damini were busy playing since luckily all of us were in the same compartment. We never realized how those three days passed; and it was the best experience that one could wish for! The train finally reached its destination, from where we were to move by road to Wellington. 'My God!', I thought; it felt as if we were in heaven. Wellington was a hill station with tea estates all around and a crystal-clear blue sky in the background, creating the most picturesque view one could ever imagine. Finally, we were at Wellington, our home for the next year, and I remember everyone looking skywards, probably thanking God Almighty for ensuring our success in the most difficult competitive examination of the Indian Army. Besides the beauty of Wellington, the Staff College was known for its finesse during reception as well as its immaculate administration. The moment we landed, we were taken straight to our already allotted house, where the kitchen was well stocked, and our luggage (which had already reached by truck a day before) was nicely laid out. It was a fully functional home before we had even reached! The meals, in case we desired, could be delivered to our home from the Officers' Mess for the next three days, the time earmarked for our settling down. I had never imagined that life could be so comfortable when you reached your new duty station since we were used to struggling for at least the first month while moving for posting to any station. I wish Wellington was a role model for every posting since the biggest challenge of military life are the frequent postings.

Our classes started in the next few days, and gradually, we got busy with our studies, the children with their schools, and the ladies were busy picking up soft skills like driving, golf, yoga and swimming. We were all aware that for the next one year, besides studies, our only responsibility was our family, and life could not have been better than what it was. Our houses were on top of a hill called the 'Gorkha Hill', and at night the view of Ooty was so fascinating; I fail to find words to describe it. In the evenings, ponies would arrive in front of each house and children would desperately wait for their pony rides, while the ladies would venture into driving around the sharp turns of Gorkha Hill. Virtually every evening, while returning from class, we would see one odd car hanging on to the trees, and on every such occasion, each one of us would be busy looking at the number plate of the hanging car. We could see each other heaving a sigh of relief, realizing that the hanging car did not belong to any one of us in our carpool. There used to be a rotational carpool between officers to go for classes, thereby sparing the balance of cars for the ladies to master their skills. There was a local instructor, Michael, available to teach the ladies. Michael was very popular among the ladies since besides his driving skills, over the years, he had mastered the art of complimenting the ladies with the most appropriate 'beauty praises', and as a result, they very much looked forward to his driving classes! All in all, Wellington was a world of its own, with small celebrations happening every evening since it always used to be somebody's birthday or marriage anniversary. On all such occasions, close friends would gather at some officer's house with loud music, which was an indication for others to join in.

Our wedding anniversary was approaching, and we decided to celebrate the evening with our friends in a hotel called Riga. It was a five-star property, and though too expensive for us, the owner, to attract the Staff College officers, used to provide a special concession for military officers. To further encourage us to head towards Riga, we were also allowed to carry our own drinks to the hotel. Therefore, many of us had started visiting the otherwise expensive property. Ooty was around twenty kilometres from Wellington, and so, Riga, with its proximity to our location, was the best option and was invariably heavily booked by the student officers on the course. The management of Riga hotel was particularly careful and decent in handling military officers as their clients, and I do not remember anybody facing any issues with regard to basic etiquettes as far as the staff at Riga was concerned. It was an arrangement of mutual needs since we found this to be the only decent place close to our college, while the hotel had to survive on regular customers from the Staff College. Since it was an anniversary party, it continued till late in the night, and since the staff were familiar with me, they had no issues hosting our late-night celebration. We had invited all our friends, and the evening would have been incomplete without Paritosh, who besides being a very close friend, was also an excellent guitar player. As the evening progressed, the demand for one more song from Paritosh continued, and so did Paritosh's demand for one more drink! Around midnight, I realized that all of us were quite tipsy, and Paritosh was particularly the happiest amongst us. Monta, Paritosh's wife, was concerned seeing

Paritosh's bouncy mood and the best I could do was to assure her that no more drinks would be served to Paritosh as we started moving for dinner, which was laid out as a buffet. I do not know how Paritosh slipped past my watchful eyes, but before we could notice his absence, we heard a loud bang! I immediately rushed towards the sound and was shocked to see Paritosh lying in a pool of blood. Luckily, he was conscious, and the manager and receptionist explained how Paritosh had boxed the decorative glass at the entrance. I could hear Paritosh trying to boast about his hard punch, and I was convinced about his strength since I had witnessed his boxing skills in the Academy. But now, the situation had become dangerous, and I immediately requested the reception for some first aid. Luckily, the injuries were superficial, and Paritosh could join us for dinner.

I left my details with the hotel to clear the bill the next morning. However, by next morning, the cost of the glass on which Paritosh had tried his boxing had multiplied many times over. Pretty soon, I realized that the hotel owner was trying to profit from the situation, and so, I decided not to pay the amount he was asking for. My particulars, that were available with the hotel authorities, were sent to our college administration, and I was summoned with a specific query, which was, 'Who was that officer who broke the glass?' It was against our principles and grooming to share names in such situations, and therefore, I decided not to share the name, but assured the college administration that I would pay the hotel owner. But now, the college administration was not interested in me making the payment, but instead, they were keen on knowing who did it. I was warned that if I decided not to share the information, I would be asked to return to my unit without completing the course. It was the talk of the town that Major Punia was being sent back from the course and my friend Paritosh could not take it anymore. Without my knowledge, he walked up to the administration and confessed that he was the person responsible for the incident. Paritosh ended up apologizing for the incident to the Commandant of the college, and that is how the situation could be salvaged. I paid the owner of Hotel Riga, who in turn became our friend, and thereafter, we were frequent visitors to Riga. Besides us, most students started venturing out to the same hotel, mainly to see the infamous glass pane boxed by Paritosh. Even today, whenever we meet Paritosh and his family, the Riga incident invariably becomes a hot topic, and on all such occasions, Paritosh has to mandatorily demonstrate to everyone how he 'boxed the Riga'!

14

BONHOMIE WITH THE INTELLIGENCE WARRIOR

Good times have an expiry date, and as was convention, our time at Wellington came to an end faster than we would have liked. It seemed like yesterday that we had arrived at Wellington in the special train, excited for a new course nestled in the blue mountains. The Staff College course felt like the college life that I had never got a chance to experience, being an ex-National Defence Academy cadet, wherein you join straight from school to get into the military. At Wellington, however, we would study hard during the day and party even harder in the evening! The proximity to Ooty was an added blessing. The year passed by in a jiffy, and now it was time to pack our bags. At the end of our course, we were presented with our master's degrees in Defence Studies from Madras University. We were told during the orientation that the Staff College was a prestigious course, and we all must expect fruitful postings at the end of the course. The day finally arrived when we marched towards our auditorium, where the postings were to be announced. The Officer-in-Charge carried a maroon file with 'Army Headquarters' embossed on it in black. That day, everybody's wife would be waiting for their husbands at the entrance of their homes in anticipation of the news of their transfer to some exciting place.

After collecting my posting order, I reached home, and there she was, my wife, standing at the gate, all curious. The perplexity on my face was an ample substitute for words. I was posted to a place that was supposedly the northern tip of our country. I was to move to Tangdhar in Jammu and Kashmir in the next fifteen days. It was a place that had no mention even on the map of India. Tangdhar seemed foreign to me as I didn't know where to start from and how to reach the location. A coursemate of mine, who had served near Tang Dhar, told me that once I reached 'Sadhana Pass' on the Shamshabari Ridge, I should sit on a jerrycan[48] and slide down my way to Tangdhar! A posting to Tangdhar

---

48 Oil container.

was an indication that our luggage would be dispatched in two different directions. I would be proceeding to my place of duty and my family would have to stay put in a separate family station. In my case, it was a high-altitude area besides being a field posting, and hence, despite my best efforts, that evening I could not cheer up my wife. We packed our iron boxes with our treasure troves and memories and reached Delhi. Luckily, my father-in-law, a retired Army officer, was working with the Oil and Natural Gas Corporation (ONGC) in Dehradun, and so, my wife moved to Dehradun along with the kids, while I headed for Jammu. We bid our final goodbye at Nizamuddin railway station and moved in different directions, separated for at least the next two years. But when you are to stay away from your family, two years feel like a million. I reached Jammu the following morning and reported at the Transit Camp. I first gave a call on my father-in-law's landline number to enquire about my wife. I heaved a sigh of relief when my son answered the phone to say, 'Papa, we have reached, when will you come?'—a question that I had no answer to, as I had not yet reported to my place of duty, let alone returning any time soon. The following day, I moved with a convoy to Srinagar, following which I started from Srinagar for Chowkibal near Kupwara, from where I was to take a convoy on the third day for Tangdhar. Normally, the convoy would drop me till Chowkibal, from where I had to join another convoy the next day to reach Tangdhar. However, I was fortunate that a vehicle, along with a quick reaction team, was sent to Chowkibal to pick me up so that I could save a day, since for the outgoing officer a day would have made a world of a difference as he would be waiting in anticipation for my arrival to be reunited with his family. As my vehicle swirled through the intertwined roads of the valley, I remembered the teary-eyed faces of my children as I had bid them adieu at the railway station. Finally, I reached Tangdhar, which was a narrow jetty into Pakistan Occupied Kashmir (POK) with an abundance of natural beauty. The name 'Tangdhar' stood for a narrow stream that used to start from the Shamshabari Ridge on the Indian side and joined the Neelam River in Pakistan. Neelam is referred to as the Kishan Ganga River on the Indian side. Luckily, it was summer, and my vehicle could get me up to our Brigade Headquarters. Otherwise, during snowfall, I would have had to walk for a day from Chowkibal to reach my headquarters. I was the Brigade Major who was to control the operations, which included firing across the Line of Control (LoC) as well as counter infiltration. It was an operationally active assignment, especially because firing across the LoC used to be an everyday affair in those days, which in turn entailed my direct involvement twenty-four hours a day, seven days of the week.

I was controlling as well as coordinating with many intelligence agencies to include both military and civil intelligence Units in and around Tangdhar to keep track of the infiltration activities. The Intelligence Bureau based in Tangdhar was headed by a gentleman named Bhakar, who was immensely dedicated towards his duty. This is in the late 1990s, when militancy in Kashmir was at its peak and this individual would stay in the town amidst the locals despite my best efforts to ensure that he stayed inside the military area. Intelligence was the backbone of all our operations, and I used to pay particular attention to look after all our intelligence warriors. While all other intelligence agencies would stay in the safe environment of our Brigade Headquarters, Bhakar had declined the comforts of our brigade saying that he would not be able to do his job in a befitting manner if he lived inside the military area. He would even move at night without any escort and all his human sources would meet at odd places well after the sun shifted west to share some of the most valuable information that he used to share with us. On most evenings, he would come to my room for dinner, and every time he moved out of our Headquarters, I would pray for his safety and bid him goodbye with a heavy heart, hoping to meet him again soon. Tangdhar was surrounded by Pakistan-dominated peaks on all sides. It was like we were living in a soup bowl. Shelling throughout the day became a part of our daily routine, and in fact, we would feel suspicious on the days that the whistling sounds of those shells were absent. Cross-border retaliatory fire would continue day in and day out. I would be on my Army telephone all through the night, checking on all forward posts and Unit locations.

One night, while I was still busy controlling the firing that was in full swing at the LoC, I got a message saying Bhakar wanted to meet me urgently. The clock showed twelve in the night, and I wondered what had got him to the camp at this odd hour. I immediately rushed to my makeshift office and saw Bhakar fidgeting on the chair. He informed me that a Pakistani girl had been rescued by the locals at a place called 'Sudpura Gap', and that the girl had been handed over to the Army post of 7 Assam Battalion. He wanted to interrogate the girl at the earliest and requested me to facilitate it. I was not aware of any such development and immediately checked the veracity of this information with the Commanding Officer of 7 Assam. The Commanding Officer shared that he had received a similar initial report and was still trying to confirm the details. Hearing this, I marvelled at the intricate intelligence network and the professional expertise of Bhakar and asked him to wait till I got further details. After a while, I got a call from Colonel Reddy, the Commanding Officer of 7 Assam. He said, 'Rajpal, I am impressed with your intelligence network, buddy. The incident is known to you even before my battalion could report it.' I thanked the Colonel but didn't divulge any further details since those who are closely associated with intelligence assimilation never disclose their source. Colonel Reddy narrated the entire incident, telling me how a Pakistani girl still in her teens had attempted suicide by jumping into a stream. Somehow, she was flushed along with the flow of water into India. Subsequently, the locals, along with some boys of 7 Assam, noticed the girl, retrieved her and resuscitated her back to her senses. Colonel Reddy further requested me to hold the information from the higher headquarters since the girl looked too naïve and innocent, because of which he wanted to push her back into Pakistan.

I got a little uncomfortable, and so decided to discuss the entire incident with our Brigade Commander, who in turn directed me to keep the Divisional Headquarters informed and asked for the girl to be brought to Tangdhar to be kept in police custody. I was immensely impressed with Bhakar, without whose information, the incident would probably have gone unreported. I assured Bhakar that he will be allowed to interrogate the girl and I also asked for a special interrogation team from our Divisional Headquarters. The girl was blindfolded and brought to the Brigade Headquarters in Tangdhar. The irony was that the LOC had divided some of the villages into two parts because of which families who were living in them sometimes ended up being on opposite sides of the border. Families that were once a stone's throw away were today divided by the line that was testimony to a bloody Partition and innumerable caskets year after year on either side of the fence. Marriages would still be solemnized across this imaginary line, but discreetly, and all celebrations in such marriages would invariably happen after a couple of days, with local police being part of the party! During the interrogation, the girl continued with her story like a broken record. She kept reiterating that she had attempted suicide since her parents wanted to forcibly marry her off to her uncle, who was an elderly person. The initial interrogation had to be carried out as early as possible, and the same was conducted by the brigade intelligence team under my supervision. On the face of it, her version of the incident appeared to be real and believable; and yet, we could not have taken any chances since Tangdhar was a highly sensitive area. She was handed over to the local police and her detailed interrogation continued for the next couple of days. We learnt much later to our horror and surprise that the girl was an Inter-Services Intelligence (ISI)[49] agent from Pakistan and was planted with a fictitious emotional story!

The primary aim of highlighting this incident was the speed with which Bhakar and his team would come to know of anything happening in Tangdhar, and this was possible as a result of the goodwill he had generated among the local people. As a result of the timely information provided by Bhakar, many such infiltration bids in our area could be averted and many infiltrators had been neutralized, as well as a sizeable recovery of weapons and ammunition were carried out. To a large extent, owing to Bhakar's personal equation with the locals and my personal relationship with him, the 'Tangdhar Brigade' slowly but surely emerged as a model for military and civil cooperation. I also recollect

---

49 The premier Pakistani intelligence agency.

how some timely and correct advice by Bhakar ensured the neutralization of an air defence gun which had made it absolutely difficult for our vehicles to move between Tangdhar and Chamkote, which was where Battalion 7 Assam was located. The Pakistanis had deployed the air defence gun in a direct firing role with a range more than that of our weapons, and as a result, any vehicle moving on the track to Chamkote would be targeted by this gun and we could not retaliate since our weapons could not reach the 'Flag Hill' where the Pakistani gun was deployed. Bhakar simply said that in case we wished to curtail this Pakistani fire, we would have to get a weapon with a longer range than the Pakistani gun. I immediately spoke to my battalion at Babina and asked them to send a missile launcher that had a range of five kilometres. After necessary letters were issued, Havildar Rattan reached Tangdhar with a detachment of Konkurs missile launchers. Initially, I deployed the launchers to observe the gun while it was firing. I had explained to Rattan that the only time he could target the gun was when the loophole of the bunker opens up to fire, which was limited to around thirty seconds at a stretch. Having made all preparations, I decided to be there on ground at the missile launcher location to target the gun, which had made our life miserable for the past month, and also caused many casualties to us. We were waiting with binoculars to see the loophole open, and the moment it opened, I screamed 'Fire'. I could immediately see a red dot (a missile as seen from a distance) approaching the gun location. The Pakistan gunner sensed the missile moving towards him, but before he could close the loophole, Rattan, with great perfection, had pushed the red dot into the bunker and we celebrated the loud bang. The Pakistani gun, along with the bunker, went flying in the air like confetti. My Brigade Commander immediately called on the radio to compliment me and confessed the advantage of having a Mechanized Infantry Brigade Major!

Thanks to this tenure where I had the opportunity to work closely with such intelligence agencies, my respect towards the professional integrity and competence of such agencies have grown in leaps and bounds. Due to my very close association with Bhakar, and having seen him work so sincerely, I ended up having an extremely high professional as well as personal regard for him. Over the years, we remained in touch with each other, and recently, I had the opportunity to attend his son's wedding. Even today, every time I meet him, we remember and cherish the 'Tangdhar bond' that we had established ages ago. I have come to realize that the more difficult the postings, the closer are the associations with people with whom you work during those trying times. The camaraderie among people who are connected with each other in such tense areas stays for a lifetime. Bhakar Saab, I salute you for your dedication, professionalism, and above all, for being my role model!

15

BONJOUR AT THE GUINEAN BORDER

High altitude, combating militants, controlling the firing at the LoC, which made even a minute's sleep a silken dream, and finally, the Kargil operations. All this made the last three years at Tangdhar, Kashmir, an eventful ride that eventually transformed my vision of the world around me, and those memories are forever etched in my journey through life. There aren't rainbows without rain, and these three years did take a heavy toll on my family life. On one side was my call of duty, the fulfilment of an oath I had made, which was to obey all commands, even to the peril of my life, as the nation comes first, always and every time. On the other side was the separation from my family, which came with its own share of agony and heartbreak. So, after endless days and nights of feeling a little blue, we as a family were really looking forward to finally living together at Babina, where my battalion, 14 Mechanized Infantry (16 Jammu and Kashmir Rifles), was located. I have always been a staunch believer in the proverb 'Man proposes, God disposes. People may come up with any number of ingenious plans, but ultimately, forces outside our control determine our future course. God's intent started unfolding on my first day at work in Babina when a signal came from the Army Headquarters, stating that I had been selected for deputation in a United Nations (UN) peacekeeping mission in Sierra Leone, West Africa. Under normal circumstances, it would have been a time to rejoice and celebrate, as all military personnel look forward to such an opportunity because of the accompanying extensive international exposure, and not to forget, a chance to draw a salary in dollars. Yet, my heart sank on reading the signal, since I could only think of my family's reaction to this news.

I had to report to Delhi to my newly assigned battalion for the mission, the 5/8 Gorkha Rifles; we were to jointly establish a new mission in the war-torn country of Sierra Leone. A cursory glance at the status of various countries around the globe immediately made it clear that Sierra

Leone was one of the poorest in the world, the poverty further exacerbated by years of civil war and military coups. The nation had also long been the victim of a rebel movement called the Revolutionary United Front (RUF), whose intentions were dubious and on the verge of barbaric. Sierra Leone appeared to be in the clutches of endless mayhem, with over 50,000 deaths, millions of people dislodged, and gruesome crimes inflicted on women, children and others, including ravishment, arson, mutilation and mass murder. Tranquility had circumvented Sierra Leone for nine years. Cities and towns were drowning in insecurity, with the supposedly vanquished rebel army indulging in malevolent retributive campaigns against the vulnerable civilian population. After years of turmoil, the Lomé Peace Agreement[50] was like a light at the end of the tunnel for the people of Sierra Leone. The agreement, signed by the RUF in July 1999, stated that the RUF was willing to lay down weapons to a neutral force of the UN. Hence, the Indian Army was given the proud privilege of setting up a mission in Sierra Leone to aid in establishing peace on alien soil. A tremendous amount of background preparation takes place at the preliminary stage prior to any mission of this scale and magnitude. For an international mission of the UN, orientation and in-depth training in

---

50 The Lomé Peace Agreement was signed on 7 July 1999 between the warring parties in the civil war that gripped Sierra Leone for almost a decade. It was named after the capital of Togo, Lomé, where the agreement took place.

the ethos of the organization were to be imbibed. Hence, the initial two months in Delhi were meant for integration and preparation, and each one of us was cautioned about the impending task. That was also when the severe routine of physical fitness, training, and the gathering of all possible information about Sierra Leone commenced. The area where we finally got deployed was the heartland where the rebel organization was most active. Colonel Martin, the self-styled Brigade Commander of the RUF, had a strong hold on the eastern province, and despite having an elected government at Freetown, the RUF was the one calling the shots. Being a peacekeeping mission, we were to ensure peace in the country and handle every turmoil created by the RUF with diplomacy and patience.

Once, on a specific request by Colonel Martin, I decided to go on a joint patrol to the Guinean border. Actually, the RUF was apprehending a Guinean incursion. I was aware that even if I tried my best to obtain official permission for a joint patrol with the RUF, it would not come through. Yet, I could smell something fishy and wanted to control the emerging volatile situation. I, as the Commander-on-Ground, had to strike a balance, and accordingly, I reported it as a routine patrolling to the Force Headquarters. As far as the RUF was concerned, I asked Colonel Martin and his team to report to my location at six in the morning. Colonel Martin reached our company base sharp at six, and we moved in two vehicles with adequate administrative supplies for the day. We were to drive up to

Lorlu village, which was around ten kilometres away, and the local Company Commander of the that area was to receive us. Thereafter, we were to move on foot through the dense thicket for another five kilometres to reach Keredu village, where the alleged incursion was reported. I drove the lead vehicle, with Colonel Martin in the front seat guiding me through the twisted roads. Enroute, we discussed how the Moa River was a natural demarcation between Guinea and Sierra Leone, thereby diminishing the chance of any border-related ambiguity. Colonel Martin pointed out the fact that at several junctions, the Moa stretched out into a number of channels and kept shifting its course. However, he did mention the border pillars, which were placed at every kilometre, and the verification of the border pillars in the disputed area would aid in indicating the actual position of the border. In an hour, after having discussed the variegated topography of Sierra Leone, we reached Lorlu, where the Company Commander received us as planned and was mighty delighted to see several soldiers in the blue berets. We started moving through the bush, which was quite dense. All of us put together were around twenty in number, a minuscule size in case the Guinean Army decided to attack. This made me wonder how we would justify our position to the Headquarters if we faced an attack, as for them we were on a 'routine patrol' around the town. Yet, here we were, with the rebel commanders, approaching an international border. We were exactly where we were not supposed to be.

While walking towards Keredu, I started discussing the further plan of action with Colonel Martin. I convinced him that a much smaller patrol consisting primarily of UN soldiers must approach the Guinean border ahead of Keredu. We decided to take along one RUF soldier, Socrates, as our interpreter, to help us communicate with the Guinean soldiers manning the border. Understanding the criticality of the situation, Colonel Martin reluctantly agreed. He decided to stay back in Keredu, which made me realize the faith this man had bestowed upon me in such a short span of time. Keredu hardly qualified as a village. It had around two to three huts, where we halted for breakfast, which was sufficiently available for everyone. Socrates could understand English and speak intermediate-level French, the official language of Guinea. At first, Socrates seemed quite agitated over the incursion by the Guinean soldiers, but I calmed him down and asked him to only do the job of an interpreter, nothing less and nothing more. After breakfast, Major Nair, my fellow Company Commander and I, accompanied by four escorts of our company, and not to forget, Socrates, started for the Guinean post. As we drew near, we could see their post at a distance. The moment we were 500 metres away from its periphery, we raised our blue berets in the air. Our hearts pounded like a train down the track, as this was our first experience of moving into an international boundary without prior notice. We were chanting our prayers and hoping that the bunker ahead of us wouldn't open fire. To make our intentions clear, we started yelling 'UN, UN.' But our attempts bore no results. As we came close to the bunker, we could only see the muzzles of their guns pointing towards us. The soldiers' deep-set eyes glared at us through the lens of the binoculars. The one-word order 'fire' would have ended our journey of peacekeeping in the depths of the interminable tropical forests of Sierra Leone. 'If they shoot us', I thought, 'would we be called infiltrators by the Guinean Army?' 'Stop', I told my mind. I had no time to indulge the pandemonium swirling in my head. I had to get my men out of this muddle at all costs. In my Academy days, my Ustaad had once told me, 'All this ragda[51] and padhai[52] might seem pointless to you now. But you never know at what juncture this learning of the Academy might save your life.' Those words didn't make sense to me then, but on that day at the Guinean border, I realized how true they were. How could I forget my alma mater, the National Defence Academy, where I had studied French? On that day, all my French expertise began to resurface in my mind. I started yelling 'Bonjour' in a high pitch, and seeing me, Major Nair as well as the other soldiers leaped on the bandwagon with screams of 'Bonjour' echoing through the jungle. I feel the bonjours did the trick for us, as a Guinean soldier 100 metres ahead of us replied with 'Bonjour,' hearing which we heaved a

---

51 Grinding.
52 Studies.

sigh of relief. The reply from the Guinean soldier was like a blessing, which immediately released our stress. The Guinean soldier smiled and said something in French that was beyond my comprehension. I guess they were cracking jokes on us, as we looked like a bunch of maniacs howling 'Bonjour' in unison, as if it was some sort of battle cry to invoke patriotic sentiments. Nevertheless, we were alive!

That was when I asked my interpreter, who hadn't been of much use till then, to inform the Guinean soldier that we were a UN peacekeeping patrol on a routine reconnaissance task. I introduced myself and Major Nair while still holding our blue berets in the air, to which the soldier smiled and said, 'Welcome.'

The Post Commander was a young officer who saluted us, offered us glasses of water, and inquired about our agenda. I informed him that I was a Company Commander of the UN peacekeeping mission at Kailahun and had approached the Guinean post to verify the border pillar on the ground. The Post Commander was a positive soul, and he started explaining the history of the border along the Moa River before showing us the border pillar on the ground. Sure enough, Socrates started confronting the Post Commander on some issue; seeing his aggressive body language, I literally had to press his hand indicating him to stay quiet. After briefing us, the Post Commander took us to the border pillar, which appeared intact and untouched, with the Guinean post aligned with the pillar. What had actually transpired was that the Moa River in that stretch had changed course that year. As a result, the border post appeared much ahead of the Moa. However, it was in line with the border pillar. Having satisfied ourselves, we thanked the young Post Commander and started heading back with a great sense of achievement. I still recollect what I had mentioned to Major Nair at the moment: 'Fortune favours the brave.' We were fortunate enough to have exchanged greetings in French with the Guinean soldiers using our presence of mind; otherwise, we would have been killed. After getting back to Keredu, I explained everything to Colonel Martin, and I could see that broad smile appearing on his face again. This was the ultimate milestone for our chemistry. Hereafter, a friendship blossomed between Colonel Martin and me, and I vividly remember committing to Colonel Martin that in case of any threat, I would be the first to take up arms and fight alongside the RUF to protect the integrity of Kailahun, my place of duty. I remember that day Colonel Martin hugged me, a hug that was warm, affectionate, and a symbol of respect from one soldier to another.

16

SOLDIERING AT THE COST OF FRIENDSHIP

The peacekeeping assignment in Sierra Leone was an eventful tenure, to say the least, with hardly a monotonous minute. This was a tenure where I was personally taken hostage by Colonel Martin despite both of us being the best of friends. In addition, my company was cordoned off with specific demands by the RUF rebels to lay down weapons. The Sierra Leone episode is a spine-chilling story of 233 Indian Army soldiers surrounded on all sides at a remote village called Kailahun. The soldiers decided to stand their ground while staring straight into the eyes of death and chose dignity over cowardice, honour over freedom, and valour over two square meals. The standoff at Kailahun continued for around three months and we were completely cut off from the rest of the world with absolutely no communication with our families and no food to sustain ourselves, and yet, our morale and motivation did not waver for a second. Eventually, having explored every possible avenue for a peaceful resolution to the standoff and with no solution in sight, we decided to fight our way back through the seventy kilometres of rebel-infested dense forest. 'Operation Khukri', launched by the Indian soldiers on 15 July 2000, happens to be one of the most successful operations carried out till date in the history of UN peacekeeping. That day the Indian soldiers, already suffering from extreme isolation and exhaustion, fought against an unknown enemy, on alien soil, a war not to capture our adversary's territory, or for that matter any lucrative or viable military objective, but only so that the tricolour could flutter with pride over the millions of hearts back home in India. My daughter Damini recounted every moment of that standoff in her own way, as a tribute to Havildar Krishan Kumar, Sena Medal (Posthumous), the only gallant warrior we lost in 'Operation Khukri'.[53] In this regard, I make a humble request to every Indian to read

---

53 R. Punia and D. Punia, *Operation Khukri: The True Story behind the Indian Army's Most Successful Mission as Part of the United Nations* (New Delhi: Penguin, 2021).

the book, *Operation Khukri*, as a real tribute to the gallant actions undertaken by the 233 Indian soldiers who did our country proud on an international platform.

After that very eventful, successful and satisfying peacekeeping assignment, I was looking forward to staying with my family after four long years of separation. In these four years, my son Arjun and my daughter Damini had suddenly grown up, and even now, I feel a tinge of pain in my heart for not being around much to witness and share in their childhood in the manner that my better half and I would have liked. I guess this feeling of guilt with respect to my absence during the growing years will always remain with me. My wife, in all those critical years during my varied field tenures, stood strong like a rock to handle every challenge single-handedly in my absence. She would play the dual role of a mother as well as a father to the kids and she did both exceedingly well as today we have two beautiful children, the credit of which goes to her as she held the fort like a true queen. This is the price that every soldier who dons the olive green has to pay, and that every family has to endure, as this is part of the sacred oath which each one of us has to take prior to serving our nation. While my wife, herself being from the same fraternity, understood the importance of such far-flung missions in field areas, it generally became extremely difficult to make our kids understand the reason behind such prolonged absence. I remember once even before I could reach home, my wife narrated an incident to me over the telephone, wherein my six-year-old angel, who came to know that I will be arriving shortly, only asked this of my wife, 'Hope Papa will not leave us now.' Hearing this, I had a lump in my throat imagining my children's inquisitive eyes, wondering whether 'Papa' will attend their birthdays and school functions. I was so excited to finally be able to go home and meet my family after such a long spell of soldiering. My anticipation was making it difficult for me to sit still at Lungi, the airport in Sierra Leone, waiting for the arrival of the UN chartered flight to take me back to India as I was aware that my wife and kids had already reached New Delhi to welcome me. As I was waiting impatiently at the airport, I was ruminating over the sequence of events over the last one year, which appeared to be too theatrical to be real. It was like yesterday when we had landed at Lungi with significantly high expectations from the locals, who were fed up with the atrocities that were being committed by the RUF. In my mind, I was satisfied and content for having lived up to their expectations because of the mere fact that we had defeated the RUF in their own bastion. It was for the first time that the world's most dangerous militia was defeated, which ultimately forced open the gateway for the RUF to return to the negotiating table. Today, Sierra Leone is a peaceful and prosperous country, and the Indian soldiers have played a huge role in restoring this hard-earned peace and tranquility. While the Sierra Leoneans do acknowledge this enormous service by our soldiers towards the well-being of their nation, the only regretful aspect to all of this is that this narrative of bravery and sacrifice by the Indian Army does not find a befitting space in the annals of our history in the manner that it should have, and most Indian citizens are oblivious to this.

While I was waiting at Lungi airport for the flight, I do not know how but word spread in Lungi town, and as a result, by the time I was about to board the flight, a thousand locals gathered to bid farewell to their beloved Kailahun Company Commander. They were acknowledging my contribution towards peace in Sierra Leone and had lined up on both sides, covering the entire stretch from the air terminal right up to the aircraft. They were singing a melody in their local dialect with gratitude in their eyes, thanking us for fighting for their freedom. I ensured that I shook hands with each one of them while walking towards the aircraft, donning black glasses to hide the moist eyes of a soldier. Deep down, my emotions were running very high, and I wondered why I was leaving Sierra Leone when the locals had started regarding me so highly. The locals were aware that the RUF's announcement to return to the negotiating table was a direct fallout of their defeat at the hands of the Indians. It was for the first time since its inception that the RUF had lost despite the best efforts of the Nigerians and other African countries in the past. Had I been given the option to stay back, I probably would not have left Sierra Leone under the circumstances. However, leaving aside my other official commitments, I was conscious of the fact that my family was eagerly waiting for me back home. I bowed my head one last time on the Sierra Leonean soil and could not hold my tears when I looked back one final time prior to boarding the flight to New Delhi. During the long flight back, I was deeply engrossed in my own thoughts, remembering Colonel Martin with whom I shared camaraderie way beyond friendship, and yet, how the two of us had to face each other on the battleground, having tried all possible peaceful options available. Colonel Martin, as a Brigade Commander, had to function directly under the orders of the RUF Field Commander, General Issa Sesay. Despite his best efforts, Martin could not convince his Force Commander to let go of the Indians along with their weapons, which would have been an honourable exit for all of us. I know for a fact that Martin even argued our case, citing the example of the Guinean patrol, when I had endangered my career for the sake of the territorial integrity of Sierra Leone. General Issa, at his end, was more than convinced about the good intentions of the Indian peacekeepers and had previously even complimented me personally during the only meeting we could have at Kailahun, as a result of the efforts of Martin. Incidentally, Major General Jetley, the Force Commander of the peacekeeping mission in Sierra Leone, could not succeed in meeting General Issa despite his best efforts. The RUF considered General Jetley to be the engineer of a Magburaka-Makeni incident in which a few RUF soldiers were killed, and it was only as a reaction to that incident that the entire peacekeeping force was surrounded by the

RUF at Kailahun. Thereafter, Martin would pull my leg every time, saying that if General Jetley had not been from India, he would have ensured our honourable exit from Kailahun.

Despite the crisis, Martin and I shared a unique bond, and as a result of our equation, we could maintain a balance despite the standoff wherein thousands of RUF soldiers had surrounded our company. I used to ask for frequent flag meetings with the RUF Brigade Commander and he would respond warmly to every such request of mine. Thanks to him, I could evacuate a soldier of mine who required urgent surgery; in fact, the RUF themselves evacuated the soldier. On our subsequent interactions, I discovered that during the standoff, Colonel Martin was aware of our depleting stock of dry rations and was playing a waiting game for the rations to get completely consumed so that we would not have any option other than to surrender. I am sure that the entire RUF was astounded when this small contingent of Indian soldiers decided to hold on, endure the hunger, and subsequently choose a fighting breakout rather than laying down arms and surrendering. Operation Khukri involved an audacious fighting breakout through the RUF cordon and an ensuing fight through the entire RUF heartland before we could marry up with the main force. The move took the entire RUF by surprise as just a day prior to launching this bold manoeuvre, I had met Colonel Martin in the garb of negotiation and had even shared a drink with him.

I still do not know whether Colonel Martin survived or sacrificed his life for the RUF since nobody I know has heard of him since 15 July 2000 ...

This is a tale of two professional soldiers, who despite being the best of friends, faced each other in the battlefield and chose duty over bonhomie. I cannot help but hold a great deal of respect for Martin in my mind. I thank him for being the person he was, as without Martin's presence in Kailahun, the sequence of events probably wouldn't have unfolded in the same manner, and ultimately, the desired peace would not have had a chance to fructify. I always pray to God Almighty that I may get an opportunity to meet my friend some day and share a drink with him one more time!

17

SANGHE SHAKTI ON BALWINDER'S FARM

In all my years of air travel, I think I have never felt the way I felt on the day when finally, after successful completion of my tenure at the UN peacekeeping mission at Sierra Leone, I landed at the New Delhi international airport. It was the most liberating flight of my life as I was flying back to my homeland with honour and dignity. There was certainly excitement in the air as I was flying back to my motherland, but I also felt nostalgic and overwhelmed to be meeting my wife and children after a long time as during the impasse it seemed as if I would breathe my last before seeing their faces. On landing, even while at the tarmac, I could see from a distance my family waiting anxiously to welcome me. I was pleasantly amused to see my wife waiting with a garland in her hand and touched to see my children holding a placard, which read, 'Welcome home, Papa.' It was an emotional reunion as we held each other in a family huddle, while I silently prayed never to be separated again. I had to continue for a month in Delhi as the Public Relations Cell of the Army Headquarters organized a number of media interactions to highlight the achievements and contributions of the Indian soldiers towards peacekeeping. Yet, unlike today's culture of 'Breaking News', none of it figured in any of the news channels. During that one-month stay, even though it amounted to my children missing school, I insisted on my wife and kids staying back in Delhi with me. The Army Day celebrations were just around the corner and an unexpected surprise was waiting for me. On the eve of Army Day, my name was approved to be conferred with the Yudh Seva Medal for my operational contributions in the face of the enemy during Operation Khukri, thereby upholding the prestige of our great nation in an international environment. It was a matter of great pride not only for me and my entire family, but also for my regiment. The next few days were spent in acknowledging the congratulatory messages I was receiving from all over the country.

It appeared that my good fortune was becoming a regular companion, as very soon, I got my new posting order in my hand. I was posted on promotion as an Instructor in the School of Armoured Warfare, Ahmednagar, Maharashtra, which is considered to be a very good peace station. And so, the next three years of my life were a blissfully satisfying time with my family wherein I could catch up on everything I had missed out on over the last four years. I had opted to wait for two years for the Command of my battalion, and it was worth the wait since the ultimate dream of every officer is to command the Battalion he was commissioned into. The same was for me, too. However, everyone advised me against the wait since it meant that even my promotion had to be delayed by two years and I would lose seniority, as compared to my coursemates. Well, I believe some decisions have to come straight from your heart and you cannot be troubling your mind every time! Finally, in April 2005, I took over the command of my battalion, 14 Mechanized Infantry (16 JAKRIF) at Karu, Ladakh. My battalion was tasked to be deployed and guard the same area where the Galwan standoff with the Chinese occurred recently and continues to make the headlines even today. Even in those days, I could not help but wonder how such a large chunk of land with an abundance of pristine beauty could be mutually accepted by both the stakeholders as 'no man's territory'. As I was commanding in that area, I always had a feeling that someday, this paradise would definitely be an ugly bone of contention between the two countries, especially since the Line of Actual Control (LAC) was perceived differently by us and the Chinese. The battalion's time at Karu went off relatively peacefully without any untoward incident, and very soon, as part of the routine moving plan of Units, our battalion was tasked to move to Patiala to be operational under the aegis of the Strike Corps. Notwithstanding the challenges posed during de-induction from a high-altitude area, our move from Ladakh to Patiala went about smoothly, and finally, the complete battalion reached Patiala. Being a peace station, all ranks of the battalion were happy and excited, as now, we would be able to stay with our families after a hard-lived field and high-altitude tenure. However, as my favourite saying goes, 'Man proposes, God disposes.' Settling down for a battalion in a new peace station can be time-consuming, and generally, at least a month is granted for the same. Yet, before we could even settle down, I was informed that our Brigade Commander wanted to meet me for an urgent task. As I entered his office, after the formal introduction and usual pleasantries, the Commander apprised me of an important forthcoming test exercise which was aimed at studying the feasibility of the deployment of the Strike Corps in the developed sector.

This was a major shift in the way the Strike Corps operations would be envisaged to manifest on ground during actual operations, and so, on my return from the Brigade Headquarters, I called for an All-Officers' Conference to explain and clarify the Commander's new directive and his intent. Once this was conveyed to my officers and the message percolated down to all relevant appointments, we initiated the process of preparation for the forthcoming exercise. Since we had just moved from the northern theatre, reorienting everyone towards the western theatre was a major challenge in itself. Operations in the developed sector would necessitate negotiating water obstacles as part of our preparation, and I requested the Commander for his permission to take my battalion for a 'flotation training exercise'. Once I explained to him the importance of gaining proficiency in crossing water obstacles, permission was granted in no time, and the battalion moved to the 'Sirhind Canal'. And so, our training for crossing water obstacles started at the Sirhind Canal. Being away for quite some time from the western theatre, the going was not very easy, as a number of our troops were new to this type of task, and every day used to be an eventful day. On one such day, while the boys from my battalion were in the process of establishing a crossing with the assistance of a safety rope, the rope suddenly snapped, and most of the boys lost their footing and balance. The situation could have very easily spiralled into an unfortunate one, as some of the boys were not very proficient in swimming. The water current was so strong that it took us over an hour to control the situation and recover the non-swimmers. Luckily, everyone was wearing life jackets. Though everyone reached the banks of the canal, all safe and sound, the mishap had a major impact on all the greenhorns that were lined up for the crossing. I could see the fear in their eyes and the hesitation in their actions. Here, ordinarily I should have lined up all of them and given a small motivational talk on the importance of training, and so on and so forth. Yet what I did next is very hard for me to explain rationally and will possibly be difficult for my readers to fathom with the correct perspective. I was fully dressed in my uniform and was standing on the bank of the canal when suddenly, without uttering a single word, I dived into the water and started swimming. It was a spur-of-the-moment decision and my whole battalion was transfixed by my actions. Mind you, neither am I a good swimmer, nor did I wait to wear the safety life jacket. Moreover, due to the fact of being fully dressed, the degree of difficulty I faced to cross the eighty-metre-wide canal had multiplied exponentially. One of my officers, Captain Kaustub Dasgupta, had been tasked as the retriever and was manning the safety boat. He immediately sprang into action, steering the boat alongside trying to pull me into the safety boat. I shrugged him off and continued swimming even more ferociously. It must have been divine intervention that despite my lack of expertise in swimming, without any life jacket and the severe water current, I managed to reach the far bank of the canal, with my entire battalion being mute spectators, wondering, 'What just happened?' Once my battalion saw me on the far bank of the canal, something magical transpired between me and all of them. Suddenly, the fear was replaced by awe, and the moment I made a signal with my hand, the entire battalion, without a speck of hesitation, jumped into the water and within seconds had crossed the obstacle. That was the evening I gave a call to my Brigade Commander and confirmed that my battalion was ready to undertake any task assigned in the forthcoming test exercise.

We mobilized for exercise 'Sanghe Shakti',[54] and as per the mobilization plan, we were concentrated in the general area around Moga and Faridkot in Punjab. Hereafter, we were to move from the western parts of Punjab towards east, as we were tasked to secure a crossing over Sirhind Canal and project ourselves in the general areas south of Ludhiana to destroy the mock enemy armour (tanks). The major challenge that was posed before us was movement with our heavy equipment through the densely populated areas of Punjab. Our brigade was to lead the advance of the Strike Corps and I was a bit relaxed since I was commanding the reserve 'Combat Group'. However, destiny had something else in the offering as I got a late-night call from our Brigade Commander, instructing me to lead the charge of the Strike Corps. I was left with hardly any time, and it was barely enough to convince and prepare my officers since all of them were understandably

---

54   Codeword.

pretty upset with this last-minute change. However, in life, I always believe in opportunities that God creates for you in the form of a difficult situation, and I thus accepted this challenge as a godesent opportunity. Just prior to our move, I got another cautionary call from the Brigade Commander, informing me that since the entire Strike Corps was to follow our battalion, I should at no cost allow the tempo of the advance to die down. So, the following morning, we offered our prayers and commenced the advance, with me skipping ahead as the leading tank of our combat group with the extremely dynamic and fluid situation firmly in control. The advance continued as per plan without a break, and on the way, we had no option but to break numerous culverts and minor bridges. We even had to put mud over the railway tracks that we crossed on our way, so that the tanks could traverse. Every field in Punjab has a farmhouse, and my biggest challenge while moving in the dark night was to ensure that there was no loss of life. Therefore, as we moved forward slowly, I understood the logic of the Commanding Officer who had declined at the last minute to lead the advance. Anyway, so far, we were moving as per the plan, and it was getting close to first light. All our tanks and equipment were occupying a viable tactical position just short of Sirhind Canal, which my battalion was to cross in a little while.

Just before we were to launch the final assault, I saw a Khalsa, with a kettle in his hand, distributing tea to all our soldiers. I was taken aback as I had not anticipated such a scenario. As he approached me, one of my boys informed me that we had just traversed through his cultivated field, and in the process, had destroyed his entire crop in the field. When he reached my tank, I enquired, and accordingly, got confirmation from the burly Khalsa that he was indeed the owner of that field. I was speechless for a moment and could not believe that the individual whose ready crop was completely razed to the ground by all our tanks was actually busy distributing tea to all my soldiers! I got down and asked for his name, to which he replied, 'Balwinder Singh.' When I asked him in an extremely apologetic manner about the loss which he had sustained, he simply told me that the Indian Army was fighting for the izzat[55] of our country, and this was the least he could do for us. He added that that he had prepared breakfast for all of us and it was ready to be served. I could do nothing other than give him a tight hug. I was emotionally charged to the hilt, feeling the colossal patriotism in his heart. Without wasting any more time, I went on the radio to declare to the Brigade Commander that the test exercise was a success, informing him that the success was partly thanks to the golden hearts of the people of Punjab, and thereafter, I narrated what Balwinder had to say.

Sanghe Shakti ended up being a huge success and I was complimented by everyone. The General Officer Commanding of our Division and the Corps personally

---

55 Pride.

complimented me for the great achievement and bestowed upon my battalion the complete credit for the success of Sanghe Shakti. Thereafter, having successfully completed the Command of my Battalion, my parting words to all my officers were to never turn down a godsent opportunity!

18

SOJOURN IN THE MECCA OF MANAGEMENT

I consider myself very fortunate to have been selected to attend the prestigious Higher Defence Management Course at the College of Defence Management, Secunderabad, Telangana. I must admit that this course was the best thing that could have happened to me personally, as well as to my military career. In fact, I can take the liberty to say that this was the best that could have happened to our family since after the completion of this course, I could hear my wife say that I was a shade better as a human being than I was before, and even my children approved of me as being a better father. It was only once the course started that I understood the real meaning of this most sought-after course, and even now, at the twilight of my military service, I wish I could have taken this course much earlier in life. I think it would be unfair on my part if I do not try and put across, being as brief as possible, the gist of learning from this 'Mecca of management' for the benefit of my readers. One of the very first lessons imparted to us was how to 'manage oneself'. It is a simple task, and yet, it is as complex as life itself, and failing to do so will generally have a negative cascading effect on everything else around us. I am grateful to my instructor for explaining it in such an easy manner that it was an eye-opener to realize the importance of self-management. It would be very juvenile to simply explain the aspect from the realms of managing actions; in actuality, it is much more than that. Self-management includes managing your emotions, relationships, routines, time, priorities, work-life balance, and above all, managing and directing your efforts for a better tomorrow. Once you learn to manage yourself, managing the world will not be as daunting as it would appear in the eyes of a naïve person. It would be much easier as the world might start getting moulded as per your requirements. I was lucky to have been imparted the first lesson of self-management in the National Defence Academy, Khadakwasla, when I was just sixteen. Even if you lost your bicycle, your senior would simply ask you to manage, and this reply was repeated to us by every senior in almost every situation, so much so that it actually became our second nature to survive under any circumstance.

Readers will recall the incident of the runaway horse, where my coursemate, in spite of being injured, managed to retrieve it. It may safely be said that management of tricky situations comes naturally to all of us in the Army and that, too, from very early in life. However, the issue explained in our management course at Secunderabad was far more than simply survival. To manage yourself, you have got to understand yourself, your emotions, your thoughts, and above all, learn not to over-react to any emerging situation that may not move in the direction of our preference. It is essential to be more balanced in life, which generally happens only after tumultuous life experiences, but the earlier you learn this, the better and happier you would be. There cannot be any set rules to live life since no two situations in life are alike. Therefore, it is essential to learn to manage yourself according to the situation, and believe me, every situation requires a different set of rules to handle it. That is why they say that management is an art and not a science, and the earlier you master this art, the better will be the quality of your life, and the happier you will be. You will never have all the solutions for the situations that may spring up in front of you as the lessons that life teaches come with age. However, one formula, despite life being an art, which would work in every situation is to 'keep it simple'. Keeping it simple is simply a philosophy and a way of life. Happiness is directly proportional to simplicity and there could be no other solution to happiness than simplicity as the route of paramount happiness passes through the town of simplicity. As I always say, the Army teaches you a way of life which no other profession can teach you, and way back while appearing for the most difficult merit-based staff college written examination, the instructor would simply ask us to 'keep it simple'. There was a written paper which was focused on solving a tactical problem and the officers who came out with the simplest plans used to undoubtedly qualify. The same goes for the plan of life, and if you decide to keep it simple, I can assure you that you will be successful and happy in life. My subordinates always ask me what I mean by keeping it simple. My mantra[56] is very uncomplicated, and that is to speak your heart, live as if it's your last day, behave like the world is glued to you, and finally, maintain an upright relationship in life. And I can assure you that you would be as straight as a ramrod and the theatricality of life would make it the simplest way to live!

Way back in 1984, I was a young platoon commander, newly commissioned from the Indian Military Academy, Dehradun. I was barely twenty years old and still trying to understand the nuances of commanding a bunch of forty odd soldiers. One fine day, immediately after the routine morning physical training parade, a sepoy of my platoon walked up to me and requested for leave since

---

56 Solution.

there had been a death in his family. I showed concern for the untoward incident but told him that I would have to ask our Company Commander for permission to grant him leave. Since we all were walking back after the parade, my immediate senior overheard my conversation with the soldier and later summoned me to his room. We used to only get half an hour to freshen up and eat breakfast, and so, despite rushing against time, I had no option but to walk up to my senior's room. He was very candid and straight in telling me that I had betrayed the confidence of my soldier, who had had a death in his family, and had walked up to me with high hopes and expectations. He further explained to me that my 'check with the Company Commander' reply would ensure that the soldiers in the future would walk up directly to the Company Commander for leave rather than approaching me. That day, my senior was trying to drive home a point that as officers, we were supposed to make decisions, and that I should have simply sanctioned his leave and informed the Company Commander later. That was the last day wherein I was caught on the wrong side of the decision-making apparatus.

Decision-making was the primary area which was reinforced time and again during our management course, and we understood exactly the steps involved in the entire process of decision-making. The first requirement is to have the intention to take a decision, and I can assure you with all my conviction that the quality of your decisions will gradually improve with time; and eventually, your decisions will gain the acceptance and respect they deserve, in the professional domain as well as in the personal space of family and home. Therefore, my advice for garnering a holistic decision-making capability is to, again, hasten slowly. Even if the situation is of such a nature that you decide to not take a decision at that particular time, well, even that is okay because the decision to not take a decision at a particular time is in itself a decision. Life is fluid and situations change by the second and so, if you decide to stall a decision for a while, it may save the day for you as well as for others. Therefore, not taking a decision deliberately at a particular moment is also a decision! It is important in life to always take a decision, the quality of which will improve with age, but the intention to do something about a situation should always be there. The military profession deals with human lives, and thereby, the importance of making timely decisions could not have been of more paramount importance, in any other profession. However, irrespective of your profession, decision-making should be a second habit, and at every split second, you would be required to take a decision. As I said, the quality of your decisions would improve with time and age but what you have to decide today is that 'I will never be found not taking decisions in life.' Everyone wants to make money, and you can earn lots of it only by making timely decisions! Look back and you will know how much money you may have lost for not being able to take a decision.

I have come across many definitions of leadership, but the one that struck me the most during the management course was all about commanding the trust of your subordinates. Being a military man, I was already more than convinced about the importance of commanding the trust of the people around you. However, post the management course, I was convinced that this is applicable in every sphere of life. We were taught various styles of leadership like the autocratic, democratic, visionary, transformational and bureaucratic; the list goes on and on. However, with my experience, today I can say that no fixed style will work all the time; you have to adapt and adopt the best possible style, depending on the nature of the situation. I can say with all my experience that when it comes to the bigger challenges of leadership, you may not face them at your workplace alone, but they might also emerge as a crisis at the privacy of your homes. You will have to steer your family through the muddy waters that a family stumbles up from time to time, and while doing so, your decisions might be questioned by every member of the family. You might end up having to explain every decision you take in the family's interest and would need decent broad shoulders to bear the repercussions as well as a large heart to accept the consequences when the outcome goes awry. In joint families, the elders of the family were generally the ones who were responsible for decisions emerging out of routine crises; and I assure you, when it came to such situations in my family, I missed my elders then, and I miss them even now. Due to their presence in the earlier days, life was simple, or should I say, their wisdom made our lives simpler. In today's' time, with the rupture in the concept of a joint family, elders are not around when you require their presence the most, and this absence is and will continue to be a huge void in all our lives. Today, when the family is constituted of the husband, wife and their children, both the husband and wife will have to constantly remind themselves of this huge responsibility of being a *role model* for their children. It is a matter of accountability for parents to groom the future citizens of this country, and therefore, while carrying on with this responsibility, it becomes essential for both the parents to brush aside their personal egos and forget about winning an argument since an unpleasant home environment often leaves a significant scar on the tender mind of the child for a lifetime. Our children look up to us, and such scenarios will not only disappoint the little ones, but they will also have an extremely adverse effect on the personality as well as the cognitive growth of children. Our leadership at home is as important as it is in our offices, and yet, maintaining a balance and isolating them from each other also assumes equal, if not more, importance. We must be careful in ensuring that office problems do not have an adverse impact at home, and vice versa. Remember, 'In a world of turmoil, a calm mind is the only sanctuary.'

Finally, I learnt the real meaning of wisdom and knowledge during the management course. Before that, for me, both the words were similar. However, today I can say that there is a world of difference between knowledge and wisdom. The day you understand this world of difference, I can assure you that the world will say that you are a wise person! Knowledge without wisdom is of no use, and the application of knowledge in your life will only be possible if you are wise. Knowledge alone does not make you wise, but wisdom teaches you to be wise, and that is why we normally say, 'He is a wise man' rather than 'He is a knowledgeable man.' Learn to be wise in life, and that will only be possible once you utilize your knowledge, otherwise you may continue to be a storehouse of knowledge without any wisdom! The Buddha spread wisdom, as compared to all other preachers who were spreading knowledge, and that is why everybody said, 'Budham Sharnam Gachhami.'[57] Life teaches you wisdom and books teach you knowledge! A teacher teaches you knowledge and a guru[58] shares wisdom. Wisdom is all about how to live life! The most knowledgeable individuals may not be wise; however, a wise person can impart knowledge to anyone even without attending school. In your life, look around for wise people to advise you in times of crisis. Wisdom cannot be learnt in a day because life is the biggest teacher of wisdom. Books may teach you knowledge, but what life teaches you is not there in any book. Learn to be wise and towards this goal,

---

57 'Let us go be in the shelter of Buddha.'
58 Teacher; preacher.

hasten slowly. In history, every ruler appointed a wise man to advise the king and a knowledgeable person to teach his children, and this, to my mind is the best example to put across the importance of wisdom!

19

GOBA THE HUNTER IN THE FANTASTIC FIFTH

Some incidents leave such an indelible mark on our lives that despite trying hard to forget them we never succeed. There are some which haunt us, some which titillate us, and some which bring tears and produce a myriad of emotions, whirlpooling in our mind and soul. Glory doesn't always come in a burst of heroism; sometimes, small triumphs and large hearts change the course of history. Come hold my hand, as I gladly take you on a distinct journey which offers enviable experiences. It all began when my family and I were celebrating my promotion to the rank of Brigadier. I was posted to take over the oldest brigade of the Indian Army, 'The Fantastic Fifth', deployed on the LAC opposite the Chinese in Arunachal Pradesh. There, I witnessed a veritable treasure trove of nature, tucked away in the north-eastern tip of India. The landscape invited me to its picturesque hills, the salubrious climate, and its simple and hospitable people. Snowy, misty, unexplored passes and tranquil lakes came together to form some of the prettiest views ever seen.

There, I felt privileged to visit the Taksing Battalion; a battalion that was not connected to any road but a walking track alongside Subansiri River took me to Taksing. Taksing always reminded me of Goba, the boy who curtailed numerous Chinese incursions with his shadow intel. And so, on my arrival at Taksing, I decided to meet Goba's mother. On meeting her, she requested me to take Goba's younger brother to be the Hunter,[59] so that he could fulfil his brother's incomplete task. The incident of Goba continuously reminded me of my duty. With great difficulty, I controlled the surge of emotions in my heart but could not utter a word. However, Goba's brother was given the opportunity and training for recruitment into the Arunachal Scouts. Prior to my moving out of Along town, I ensured that even Goba's son was enrolled into the newly

---

59 Term coined for selected locals who worked as the eyes and ears of the Army around LAC, providing intel about Chinese incursions.

raised battalion of the Arunachal Scouts. It was the most memorable moment I cherish even today, seeing young Goba in a military uniform! Tears rolled down my eyes when he saluted me and wished Jai Hind[60] in the typical Goba style. Goba was barely sixteen years old when I first saw him. Let me take my readers back in time and tell you the story of this bravest of brave boy. Before meeting Goba, God was kind to bestow upon me the honour of taking over a brigade which had proved its mettle in the theatres of North Africa, Eritrea, Syria, Italy and Greece during the Second World War. I boarded a flight from New Delhi for Dibrugarh, and after landing at Dibrugarh, I experienced the most exciting challenge of crossing the Brahmaputra River in a huge ferry, which could even take my vehicle on board. Back then, Dibrugarh to Likabali was not connected by a bridge, and therefore, the only option available was a ferry. Finally, having been driven on a vehicle for around eight hours on a swirly jungle track, I experienced one of the most picturesque drives ever. A few hours, and two transit camp halts later, I reached Along town, which is often called 'Aalo' by the locals. Initially, I was dumbstruck by the natural beauty offered by Aalo on the banks of the Siyom River with trees dancing, choreographed by the eastern winds. I got even more excited when my officers briefed me about the alluring pristine beauty of the areas ahead, towards the LAC. Considering the vast area of our

---

60   Long live India!

brigade, a helicopter of the Indian Air Force was located at Along and I was the only Brigade Commander of the Indian Army who had the authority to utilize this asset at my discretion.

'What an amazing assignment', was my first thought on assuming the command of the brigade, and I thanked God Almighty for his blessings, without which a Mechanized Infantry Officer could never have taken over the command of a mountain brigade that was strategically so important as it looked after 450 kilometres of the LAC opposite the Chinese. The LAC is the demarcation line between Indian and China and is also known as the McMahon line after its negotiator Sir Henry McMahon. I noticed that wherever the McMahon Line drawn by Sir Henry McMahon during the 1914 Shimla Convention followed the 'Watershed Principle', there were no dispute over territory between the Chinese and us. Here, let me explain the Watershed Principle; the alignment of the highest ridge line running from east to west was like a watershed between the two countries, and its alignment was a good border demarcation, with no confusion. However, wherever the Watershed Principle was deviated from, it resulted in a perceptible difference in the McMahon line. The McMahon line was the LAC between China and India and today there exists disputed areas between India and China as Sir McMahon deviated from the watershed principle in more places than one. The logical question that comes to mind is: Why did he deviate from that ridge line? The answer is that Sir Henry McMahon did so to accommodate religious sentiments, since in certain areas like the 'Potrang Lake', the religious belief of Tibetans advocated that going around the lake once in a lifetime was a must for every Tibetan to ensure that they go to heaven. Therefore, to include 'Potrang Lake' in Chinese territory, McMahon deviated from the watershed principle. As a result, with time, many disputed areas came into prominence and even now there is no clarity about the ownership of these disputed areas.

Based on the Sino-Indian agreements of 1993 and 1996, both sides can patrol the disputed areas. However, no permanent presence or structures are permitted. The disputed areas opposite our brigade were frequently patrolled by both sides, but I was keen on continuous surveillance, and that was only possible with boots on the ground. Since both sides were prohibited from maintaining a permanent presence in the disputed areas, the only option left was frequent patrolling of those areas. I firmly believed that under the circumstances, the only option left was to fall back on the locals, and luckily the local tribals called 'Tagins' were natural hunters. Tagins were offensive by nature, well-built, and could survive off the land, being born hunters. Therefore, I could set the ball rolling for the selection of contractual hiring of locals to be deployed to the disputed areas for continuous surveillance. I also coined the term 'Hunter' as a designation for the selected locals since they were to be deployed in the pretext of carrying out hunting for their livelihood. I personally monitored the selection process and was particularly impressed by a young boy named Goba. Goba was short in stature but stoutly built, He was barely sixteen at the time. The locals were all praise for him and narrated an incident where he had killed a leopard at the age of twelve. His father was deceased, and he was the sole breadwinner of his family, with a number of younger siblings to look after. Goba would play with the most poisonous snakes found in that area and in his village that was called 'Tame Chu Chu' meaning the 'Land of Snakes'.

Besides the many skills that Goba possessed, I was particularly impressed to see him fishing with a spear in the flowing waters of Subansari River. He could speak Hindi well besides the local dialect, and every time I visited Taksing, the Battalion Headquarters of 23 Rajput, I would look forward to meeting Goba. He was deployed in the disputed area of 'Bisa-Maja', which was the most frequented area by the Chinese, and in no time, had qualified to be our official eyes and ears in that sector. He was presented a small radio set and a pair of binoculars by the Commanding Officer so that he could communicate timely information to the Army every time the Chinese would venture into the 'Bisa-Maja' area. He would live off the land in his area of deployment, while his family was kept happy receiving a timely payment of salary from the Army. Goba would deploy himself right on the mountain pass, which happened to be the highest point on the LAC in that area, and so was in a position to observe the Chinese right up till their base with the help of the binoculars. This arrangement worked very well for us since the moment the Chinese commenced movement from their base, we had the information, and our patrol would reach the disputed territory even before the Chinese could come into the area. The Chinese, on their part, used to be extremely surprised to see an Indian patrol in the area every time they ventured into it. The idea of using 'Hunters' for real-time information was immensely appreciated by General Kulkarni, our Divisional Commander, and he asked all other Brigade Commanders to replicate the innovative idea in their respective brigades.

Goba had integrated so well with 23 Rajput deployed in that area that every visitor to their battalion had to meet him, and on such occasions, the Commanding Officer would list out all his achievements in front of the visiting dignitary. Gradually, Goba gained tremendous popularity in the entire Calcutta-based Eastern Command, and I recollect receiving a personal call from the Army Commander to compliment me for this most outstanding and groundbreaking idea! Every time I visited the forward areas, I made sure to meet Goba, who in turn would smartly salute me, attired in his traditional dress but never without his Indian Army combat cap. A loud 'Jai Hind!' with a wide welcoming smile was part of Goba's greetings every time I met him. He would also prepare a special lunch for me, which included rice cooked in banana leaves, and trust me, I have never relished a meal more than the one offered by Goba in the interminable forests of Arunachl; a place so remote that it was as if we were transported to the Stone Age! During one such interaction, I learnt that Goba's grandfather had sacrificed his life fighting the Chinese alongside the Indian soldiers in the 1962 War. With rage painted on his face, Goba expressed his hatred towards the Chinese and assured me that any nefarious designs by them would be crushed by him alone as he would be the wall that the Chinese would have to bulldoze to reach India. Every time the Chinese patrol would enter his area, he would shadow them discreetly throughout their stay in the disputed area. He would pass important intel regarding the number of Chinese soldiers and types of weapons carried by them to our forward post. With the area being dense with foliage, it was not much of a problem for Goba to carry out this shadow movement following the footsteps of the Chinese. For the brilliant work that Goba was executing, in addition to his salary, I would also ensure that I looked after his family by providing all types of administrative support. Once, I recollect, that his mother was unwell, and soon her condition worsened. I ensured her air evacuation to the nearest government hospital. Apropos, such assistance only enhanced Goba's dedication towards the operational responsibility given to him. Very soon, he had integrated with our forward posts in a seamless manner and would visit them once a week to pick up his essential rations and other administrative support.

The idea of employing hunters for real-time surveillance of the Chinese patrols was slowly getting its due recognition right up to the Army Headquarters, and from time to time, on corroboration of the intelligence being provided by Goba and other such Hunters, a number of nefarious attempts on part of the Chinese could be foiled.

Everything was going well until, one night, I got a distress call from the Commanding Officer of Taksing Battalion, informing me that Goba had gone missing during one of his shadow patrolling missions. I was jolted to hear the news. The initial report of a Chinese patrol entering the Bisa area was passed on by Goba, but thereafter, he had gone out of communication. We launched a patrol immediately to scan and search the entire area, but in the process, could only trace Goba's radio set with no signs of him anywhere. The next morning, I instructed the Commanding Officer to carry out a detailed search of the area right up to the LAC. However, despite our best efforts, Goba remained untraceable, and since he occupied a very soft corner in my heart, I remember leaving no stone unturned in search of him. It was getting more and more painful for me to digest the fact that Goba could be gone forever and that there was no trace of him anywhere. I was heartbroken and felt as if I had lost my own child; deep down, the regret of employing Goba as a Hunter was killing me. A few months after his disappearance, a Hunter informed us that he had seen Goba accompanying the Chinese patrol. Having interacted with Goba a number of times and knowing his background, I personally did not trust this report because as far as I knew, this would be the last thing he would do. Goba's disappearance is a mystery which remains unresolved till today, and it haunts me time and again as I keep thinking about him. I still feel the guilt deep down of launching a teenager into a situation which was actually our responsibility!

20
SCOUT CAMP AT FIFTY ON AN ISLAND

When I was commanding a brigade in Arunachal Pradesh on the LAC, owing to its strategic location, the Divisional Commander would often ask me how his division was faring. To every such query, I would crisply reply that it was in 'safe hands, Sir!' The reason was that since the Division was newly raised, it was yet to take over the operational area, and for all practical purposes, I was the one commanding the Division on ground, while the Divisional Commander was seated hundreds of kilometres away at Zakhama in

Nagaland. Years later, I was offered command of the same Division, to which I had replied that there would be no challenge in commanding the same Division again. The Military Secretary was taken aback and asked me when I had commanded the Formidable Fifty-Six Division. I explained how the command of the Fantastic Fifth Brigade had incidentally also meant command of the Formidable Fifty-Six Division. A few years later, based on the directive to increase the boots on ground along the LAC, the area which was earlier commanded by a brigade was taken over by a Division. In life, many a times, your progress is decided by destiny. As a Brigade Commander, I could have asked for nothing more, since I was commanding one of the most challenging areas in the eastern theatre, which incidentally happened to be the only brigade of our Division which was operationally committed, while the other two brigades were still undergoing their establishment.

The National Defence College course is one of the premier and most prestigious courses in our country, offered to officers as per their respective merit post the successful completion of their command of a brigade. I shall remain indebted to the Fantastic Fifth for ensuring my nomination for the National Defence College, that, too, for getting selected to undergo this particular course at Bangkok in Thailand. As a family, we could not have asked for more, and my wife and children were thrilled to travel to such an exotic country. The following year at Thailand was an amazing experience, in addition to the immense learning alongside the elite of the government and the eminent industrialists of Thailand. This course in Thailand is attended by senior military officers, other senior government officers and selected business community representatives who, in particular, pay for their course expenditure. The total strength of the course was 300, including foreign officers from various countries friendly to Thailand. The first day in the college took me back to college days, about which I had only heard stories, since I had missed out as a result of joining the National Defence Academy straight after my school. While there were boys and girls all around, most of us were fifty plus. However, each one of us in the initial days ended up behaving like teenagers in college. I was particularly thrilled to meet officers from such diverse backgrounds and countries. Though conversing was a bit of an impediment, it surely didn't come in the way of me having a wonderful time in the college.

The National Defence College was the official think tank of Thailand on matters regarding strategic planning, and the national security strategy of the country would come out of this organization. The Prime Minister would come to the college twice during the course, not to address the students, but to simply make his notes while attending the presentation on national strategy fabricated by the students. I am highlighting this fact to put across the importance rendered by Thailand to their national Defence College because on the contrary, students who would have attended the same course in India would agree with me that the potential of our own national defence colleges are grossly undervalued as well as underutilized in terms of the provision of value additions with respect to the national strategic direction. The course involved frequent travelling as part of our official tours, both within Thailand as well as to other countries. The first such outing happened to be a 'scout camp' on a remote island where we all were to live in tents on the beach. The aim of the first outdoor camp was to ensure that everyone knew one another. We were all really excited about the camp since we were instructed to get our scout uniforms stitched, i.e., khaki shirts and shorts stitched, for which expert tailors were called to the college. There was a lot of excitement in the air since we were to live on a beach for five days and follow the routine of a scout. The day we learnt about the forthcoming scout camp, we all started planning in our heads. What was going on in my head was the fact that if I had to be a scout, I ought to get a backpack. Yes, a backpack was my primary concern at that moment. The very next day I visited the famous Chatuchak Market and was blinded by the bling and the variegated things available in the market. I'm not kidding! All I could find was a shiny silver holographic backpack. It was way out of my comfort zone but as they say, when in Rome, do as the Romans do. The scout camp was quite similar to a National Cadet Corps camp in India, with the only major difference being the age at which we were to undergo this camp. On the day of departure for our camp, we were to report to the college at 5 a.m. I reached the gate wearing khaki overalls and a fedora hat but not without my showstopper: the silver backpack. Seeing all of us in our shorts and scout hats were sight difficult to explain in words. We were having a hearty laugh just by looking at each other. What impressed me the most during the camp was the meticulous planning displayed by the college. We were 300, and yet, every detail, from the seating plan in the buses to baggage tags and packed meals, were so well coordinated that there was no question of any confusion. Finally, the buses started for an unknown destination, and during the journey everyone had to introduce themselves. It was time for me to put to trial my vocational skills in the Thai language. The moment I said 'Swasdee Khap' ['Hello' in Thai], everyone was in splits because of my accent. You can imagine how a Jat from India would sound speaking Thai amidst the locals would sound and hence I don't blame them. Here I should mention that because the entire course was to be conducted in their national language, I had undergone a pre-course in Thai language at New Delhi.

The moment we reached our campsite, we were all put through our first activity, which was to pitch our tents as part of our living accommodation. The view, with the pristine white sand on the beach and the clear blue sea, was mesmerizing. I could only thank the Amighty and some of my good deeds to have earned a course like this. We were allotted various appointments and were to be self-contained for the next five days for which rations were issued to us and we were to cook our own meals. Every 'scout group' had an instructor who would guide us through the various activities. Come evening, the cultural programme would start with a bonfire in the centre. I can now say with much conviction in hindsight that they were the best five days of my life as we were quite literally reliving our childhood. I compliment the Commandant, National Defence College, for planning such an unique activity as part of their course curriculum. Another aspect included as part of the overall course schedule was the guest lectures being delivered by select professionals from varied fields would deliver guest lectures, and the most interesting part of such lectures used to be the question-and-answer sessions post the lecture. There was a plethora of experience and knowledge among the students as a result of representation from all walks of life. Therefore, the biggest learning on this course was more lateral, as we were learning from each other's experiences. While the schedule for foreign study tours was being worked out, I managed to convince the college authorities to include India as part of the programme as well. I was heading the first such study group of the National Defence College students from Thailand to India. The course was nearing its end, and the mere thought of it was disturbing our peace of mind as a result of a solid bonhomie, which had developed among the students over the past few months. Towards the end of the course, we all were to present a research paper, and my paper was selected to be presented to the Prime Minister of Thailand during his visit to the college. I had carried out extensive research for finding an alternative to the Strait of Malacca, which connects the Indian Ocean to the Pacific Ocean. It is one of the most important shipping lanes in the world with very heavy traffic. After listening to my presentation, General Prayut Chan-o-cha, the Prime Minister of Thailand, walked up to the podium to congratulate me and asked for a copy of my research paper. I learnt later from my fellow students that my paper was included in the national strategy of Thailand. The southern part of Thailand is very narrow and it is actually the shortest approach from the Indian Ocean to the Pacific. I had recommended a fill or a cut route through the land mass for commercial vessels, which would, besides generating a large revenue for Thailand, also be a big relief on the heavy shipping traffic through the Strait of Malacca. I am still in touch with most of the students from that course and I was delighted to learn that my research is very close to seeing the light of day. It would indeed be very satisfying since small ideas can develop into big ventures, and such out-of-the-box ideas shatter the glass ceiling and can have a profound effect on the geopolitical milieu of the world. My friends from Thailand tell me that my name is still attached with this project like a copyright, and someday, you never know, there may be a 'Punia Strait' in Thailand connecting the Indian Ocean to the Pacific!

Rejoice in Adversity, Triumph in War

The most important lesson I learnt on this course was, unexpectedly, about Indian culture. I was amazed to hear from my fellow students that they owe their culture to India, and whatever I had seen as a child in India was also being practiced in Thailand. Even the Prime Minister would remove his shoes prior to entering anyone's home and it used to be a common sight to see young people offering their seats to the elderly in the metro as well as in other public transport. A handshake was not part of their culture as they would simply join their palms in a namaste.[61] The most amazing experience was to visit Ayodhya in Thailand, which they call 'Ayutthaya', and which is a true replication of the city of Lord Rama.[62] The historic city of Ayutthaya was founded in 1350 AD, and it used to be the capital of Thailand. Even the King of Thailand adopted the title of Lord Rama. Unimaginable in the wildest of our dreams, even a single visit to Ayutthaya in Thailand would be a grand showcase of the ancient Ayodhya to the tourists. However, none of the Indian tourists are aware of this resemblance between Ayodhya and Ayutthaya, which is why Ayutthaya is often missing from the map of Indian tourists when they visit Thailand. In Thai, the official name of the new capital at Bangkok retains Ayutthaya as part of its formal title. The ruins of the old city now form the Ayutthaya Historical Park,

---

61   The method of greeting in India.
62   An Indian mythological figure and king.

an archaeological site which has also been dedicated as a UN Educational, Scientific and Cultural Organization (UNESCO) World Heritage Site.

Coming back to witnessing the Indian culture in Thailand, the Bangkok International Airport is called 'suvarnabhumi', a word picked up from Sanskrit, meaning the 'golden land'. Millions of Indian tourists visit Thailand every year, but I can assure you that not even one would have noticed this while landing at that airport. Incidentally, the Thai language also originated from Sanskrit and the same is acknowledged by the people of Thailand. The Thai people also acknowledge the stamp of Indian culture on their customs and traditions, and I would even say that they thank India for sharing their culture with Thailand. Unfortunately, Indians themselves have forgotten their own culture, but I was happy to see our culture still alive and being practiced by the people of Thailand. I thank God Almighty for giving me the opportunity to spend one year in Thailand, though of course it took me a lot of time and effort to get adjusted again to the new Indian culture once I got back home! Deep down, I believe that we in India need to introspect as to where are we heading! Development and progress in India have come at the cost of our culture, and we need not blame anyone else for this so-called modern India with zilch values. We owe it to our future generations to pass on the original Indian culture to our children, and towards this end, let us resolve to hasten slowly and keep our feet glued to the ground while our hair oscillates with the winds!

21

THE SOLDIER WHO VACATED DERA SACHA SAUDA

26 August 2017 was an unremarkable day of business in the office. I had just returned from Jaipur after a terrific performance by our Division in the Command War Game and was still basking in the glory of the most innovative operational plan presented by our Division, which was well appreciated by everyone, and the Army Commander personally complimented me for presenting the most ingenious plan. Being co-located in Hisar, Haryana, with the Jindal Group of Industries, we were honoured to have received the high

mast national flag by the Naveen Jindal Flag Foundation of India, which was now fluttering outside my office. The glass bay window in my office had the most spectacular view of the national flag against the backdrop of my Division insignia and lush green lawns on either side. The General Officer Commanding's office was on the first floor, and as the front wall was structured of only glass, the national tricolour adorning the emerald background was the most coveted view, which was a constant delight and a matter of great pride. It was an easy day without any pressure, and having just arrived at the office, I was organizing myself for the day, when the sudden buzz of the telephone broke the silence. The signal operator of our Military Exchange was exhilarated to inform me that the Chief Minister of Haryana was on the line. I asked him to put me through and it happened to be the Additional Secretary to the Chief Minister, who in his typical Haryanvi accent, promptly informed me, 'General Saab, CM Saab aapse baat Karna chahate hain.' ['General Officer, the Chief Minister would like to speak to you.'] Even before I could utter a word, I heard Shri Manohar Lal Khattar telling me that the situation in Sirsa would require the intervention of the Army to assist in the maintenance of law and order, post the conviction of Sant Gurmeet Ram Rahim.[63] I asked the Hon'ble Chief Minister how soon we would be required, and he said at the earliest. He wished me good luck, and I

---

63   The leader of Dera Sacha Sauda, a religious cult.

assured him of our best efforts. The series of actions that followed included immediately asking my staff to issue a warning order for the movement of Internal Security Columns to Sirsa and organizing an urgent conference of all Brigade Commanders. I personally called up our Corps Commander and informed him of the development, and he wished me success and assured me of the commitment to any assistance from the Corps Headquarters. After that, I asked my exchange to put me through to the Deputy Commissioner, Sirsa, so that I could obtain first-hand information from ground zero. I also called for my Colonel General Staff to immediately dispatch a Liaison Officer to Sirsa, who would act as a bridge between our formation and Sirsa administration.

Prabhjot Singh, Deputy Commissioner, Sirsa, highlighted during our conversation the importance of implementing the court order to vacate the Dera Sacha Sauda, but he was not hopeful about the followers leaving the Dera at their own discretion. While passing the judgment for the conviction of Sant Gurmit Ram Rahim, the court had directed the administration to take over the Dera premises. The administration had to implement the court order, and therefore, it was imperative to vacate the Dera, as also to maintain the law-and-order situation. By the time our conference commenced, I was primed with a fair amount of vital information about the situation prevailing in Sirsa, and accordingly, briefed all my Brigade Commanders. My briefing was followed by a quick brainstorming session to decide on a comprehensive plan. Our task was to primarily re-establish the rule of law in Sirsa town and assist the administration in vacating the Dera. As a result of our initial review of the situation, I considered it to be prudent to mobilize one brigade to establish a cordon around the Dera. Also, in the process of building the cordon to restore law and order, an exhibition of the force was displayed by way of flag marches and in case the situation dictated it, the use of force was the last resort. I wanted to keep the balance of the Division as a reserve for the physical clearance of 'Dera Sacha Sauda', which was enormous in size, and over the years had developed into a city of its own.

The preliminary analysis of our task clearly indicated the importance of reaching Sirsa on time as our top priority. Luckily, the warning order for mobilization had been immediately issued the moment I had received the Chief Minister's telephone call. Also, a few months before our formation was deployed for the restoration of law and order during the 'Jat Agitation', and so the drills were fairly well rehearsed; and the boys knew exactly what their role was. I was aware of the mayhem and chaos created by the Dera followers in Panchkula the previous day, immediately after the judgment was passed. Therefore, I warned all my Brigade Commanders to remain focused and not take the situation lightly. The moment the conference got over; I gave the executive order to my Brigade Commander to move along with his brigade to Sirsa. Sirsa was just about a two hours' drive for a single vehicle, and considering the size of a Brigade's Convoy, I gave him five hours to be effective in Sirsa. Thus, as the conference had finished at around 11 a.m., I expected him to be effective in Sirsa by 4 p.m. The moment the meeting was over, I got the Deputy Commissioner's call, informing me of six civilian casualties as a result of firing by the paramilitary force. A mob had attacked the paramilitary, and there was no option but to fire. He further requested me to ensure that we reach as early as possible. I understood the gravity of the situation and decided to personally reach Sirsa to take control at the earliest and asked my staff to inform the convoy to reach Sirsa as early as possible. I was fairly conversant with Sirsa, having been deployed there for flood relief in 1993 as a young Company Commander. My battalion was located at Hisar, and that year, as a result of intense rainfall, there were floods in Sirsa due to the Ghaggar River overflowing, with water even entering Sirsa town. I distinctly remember that the same Dera Sacha Sauda had catered for our food while we were busy saving precious lives in the flood situation. The Dera had been established in 1948 at Sirsa by Mastana Balochistani. He was born in 1891 in Kalat, Balochistan, which was then a part of undivided India, and is now in Pakistan. He was popularly known among his devotees as His Holiness Beparawah Mastana Ji Maharaj. The Dera originated as a social welfare and spiritual organization that advocated humanitarian and selfless service to everyone and had followers from every religion as its disciples. With devotion and hard work, the followers of Dera transmuted the barren land of Sirsa into a spiritual oasis, and their Guruji imparted the great practice of meditation to his followers. The current Godman, Sant Gurmeet Ram Rahim Singh Insan, took over the divine apostleship of the Dera on 23 September 1990. The present master was initially instrumental in leading the humanitarian endeavours, and under his leadership, the Dera scaled glorious heights. The Dera was also running three specialty hospitals in Sirsa, where health services were provided free of cost to poor people. In the year 1993, during the flood relief operations in Sirsa, I witnessed the noble work carried out by the Dera and was a witness to the mass following that the current Guru Gurmeet Ram Rahim enjoyed. However, subsequently, the rape charges against the Gurmeet Ram Rahim had shocked millions of his followers spread across several states with a large concentration in the states of Punjab, Haryana and Himachal Pradesh. Finally, the special Central Bureau of Investigation court at Panchkula convicted him of being guilty of the rape charges against him. Accordingly, he was arrested, and the same court ordered the Dera premises to be vacated and taken over by the administration.

Sirsa city is located in the western-most region of the Indian state of Haryana, bordering Punjab and Rajasthan. Its history dates back to the time of Mahabharata. At one point in time, the mythical Saraswati River is said to have flowed in this area, and the name Sirsa has its origin from the sacred river, Saraswati. During the medieval period, the town was known as Sairishaka. The material remains of an ancient fort can still be seen in the south-east of the modern city, and the circumference of the town is around a few kilometres. It was one of the most important fourteenth-century towns in northern India. Sirsa was located on the ancient route leading to Taxila and was believed to be one of the oldest towns in Haryana. The city is situated on the edge of the great Indian Thar Desert.

While I was busy collecting all the possible details about Sirsa, I got a call from my Brigade Commander, informing me about his reaching Sirsa an hour before the designated time. He also shared with me that the entire city was under Section 144[64] of the Indian Penal Code. I instructed him to establish his administrative base at the Sirsa Air Force Station and get on with the designated task and wished him good luck and Godspeed. Though Sirsa was not very far, I considered flying down more appropriate in order to effect aerial reconnaissance. This way, I could easily understand the layout and extent of the Dera since I was going to Sirsa after a long gap of nearly twenty-four years. While airborne, I could discern the city being ablaze, with smoke exiting from several areas in the city. The Dera was an extension of the township, and it was like a massive city within the city. From above, I could understand it would not be a simple exercise to put a cordon around it, as the western periphery of the cordon would be

---

64 Gatherings of more than four people are not allowed.

sandwiched between the Dera and the city. While fly over the Dera I saw hundreds of people inside, and the pilot, sensing the gravity of the situation, requested me to avoid a second pass over the Dera. Finally, I landed at the Sirsa Air Force Station, where I was received by my Brigade Commander. He briefed me the moment I landed, and we immediately drove down to the Ad hoc Operations room where I met the young Deputy Commissioner, the Superintendent of Police and other officials, including the paramilitary officers. The Deputy Commissioner gave me a preliminary briefing on the prevailing situation, and post that, I considered it to be more appropriate to go on the ground. I had boarded my vehicle and was about to start, when I saw the Deputy Commissioner rushing out of the Operations Room as he requested for a conversation with me in private. The Deputy Commissioner shared with me that he had been on a call with the Hon'ble Chief Minister while I was moving out and intimated that Shri Manohar Lal Khattar was monitoring the situation personally since it did not augur well at Panchkula. He was discreet in sharing with me that in case the situation was not handled well at Sirsa, the Chief Minister was facing the threat of being sacked. I assured him of our best efforts, irrespective of someone's chair being threatened. I amplified in clear terms the fact that the Indian Army was there to do its job in the best possible manner in every situation. In any case, it is our duty to restore law and order whenever requisitioned by the civil administration. I remember noticing a smile on the young Deputy Commissioner when I asked him to inform the Hon'ble Chief Minister to sit comfortably on his chair since it was in the firm grip of the Indian Army.

From the time our troops had reached Sirsa, the Indian media, including all English and Hindi channels, were flashing misleading headlines: 'The Indian Army has entered the Dera' was exhibited on every television screen. This particular news item was providing tremendous television rating points (TRP) to channels, and the nation was convinced about what the media was showing. In reality, only a couple of hours had passed since we had arrived, and let alone entering the Dera, we had not even put the cordon around it. The most urgent concern was to restore the law-and-order situation in Sirsa before thinking about the cordon. The constant news, or false propaganda, about the Army entering the Dera was adding fuel to the flame in Sirsa, and as a result, there were people in the city wanting to even self- immolate. Therefore, the immediate cause of concern was to arrest this false propaganda as it was turning out to be an operational necessity. My Brigade Commander and I contemplated various options available before us, and I, ultimately decided to take the bull by its horns. I inquired from the Deputy Commissioner about the number of reporters present in Sirsa and the channels being represented. I was happy to get this response from him, 'You name a channel, and the reporter will be in Sirsa.' I realized that the only option available under the prevailing circumstance was to organize a press brief to personally throw light on the scenario. I was well aware of the standard procedure and protocol attached to such media briefings, wherein you have to prepare a press release and forward the same through the bureaucracy and await a formal consent from the Army Headquarters. Time was a constraint since the situation was escalating, with repetitive news stories flashed on the channels. I thought for a moment and remembered the notable teaching in our schools of instructions that when in doubt, fall back to the 'man on ground'. Suddenly, the man on the ground took control of the situation and I asked the Deputy Commissioner to arrange for an urgent media briefing.

The media personnel were amazed by this sudden development since, for the past twenty-four hours, not a soul was ready to interact with them, and yet, every channel was showing live visuals from Sirsa incessantly. I started the media brief with the initial news of an excellent example of coordination and joint-man-ship showcased between the Army, civil administration, and the paramilitary forces. Thereafter, I came straight to the point and clarified that the Army had no immediate plans as yet to enter the Dera. Furthermore, I clarified the misinterpretation because of which the channels were showing that 'the Army has already entered the Dera.' After that, I took a couple of questions in which I reinforced again that there were 'no immediate plans as yet to enter the Dera'. It was a classic example of utilizing the strength of the media to communicate a message to the masses in the fastest possible manner. We could see the immediate transformation in the headlines shown by all media channels from their earlier stance of the 'Army has entered the Dera' to the 'Army has no immediate plans to enter the Dera.' The news headlines and my interview were continuously shown by all English and Hindi channels for the next twenty-four hours since Sirsa was the focus for everyone as a fallout of previous day's events in which the Dera followers had ransacked Panchkula city. Moreover, I had deliberately chosen to give my interview in Hindi to reach out to a large number of followers of the Dera Sacha Sauda. I started getting frantic calls from our higher Headquarters, questioning me for not taking clearance from the Army Headquarters before going ahead for the media briefing. My response to all such queries was that it was an operational necessity, and the decision was taken by the man on the ground.

The biggest concern bothering me was the importance of avoiding a situation of tackling two fronts simultaneously. In case the cordon around the Dera was executed without addressing the scorching law-and-order situation in Sirsa, it may have resulted in our troops getting squeezed between the city and the Dera. The Dera and Sirsa were contiguous to each other, and therefore, two fronts would have definitely hurt the chances of our success. Military strategic thinking also advocates that such a situation ought to be avoided, as it would be catastrophic for any force in combat. Therefore, I was trying to restore public order in the city before establishing the cordon around the

Dera. I was eagerly waiting for a positive response to my media statement, and sure enough, I started getting positive feedback from the ground. The people protesting on the streets of Sirsa had started returning to their homes after my assurance of the Army not having any immediate plans to enter the Dera. I was privy to the local following enjoyed by the Dera in Sirsa since my first experience during the flood relief operation in 1993. My media statement was also targeted to pacify the locals in Sirsa so that we could effectively proceed with our task to establish a cordon around the Dera. In today's world, psychological warfare assumes tremendous importance, and it is an extension of military munitions. There is no denying the fact that there could have been no other agency more potent than the media to be our source of psychological warfare in that situation. What one statement of mine achieved, probably a million rounds of fire would have failed to accomplish. I could sense the stage being set, and it was now the right opportunity for me to give the go-ahead to my soldiers to enforce the cordon around the Dera. Our troops gradually moved to the designated locations around the Dera while the paramilitary and Haryana police continued to tackle the law- and-order situation in Sirsa.

The Dera was like a city in itself and effecting a cordon around such an extensive area was not an easy task by any measure. The only option available to blanket such a large area was by going linear without holding any inherent reserves in the cordon. This, in any case, was not a cause of concern for me since I had catered for adequate reserves from my Division. The Armoured Division that I was commanding was adequately trained in the isolation and cordoning of any significantly sized built-up area; this happened to be one of our primary tasks. However, the only difference from the current situation was that we usually did such tasks with our primary equipment, i.e., tanks, infantry combat vehicles, artillery guns, and here we all were dismounted and on foot. The Indian Army gives you enough training for flexibility in operations, and thus, it was not a significant cause of concern. What typically follows in the sequence of operations is isolation, followed by a cordon, and finally, its capture. The third stage of capture in the current situation was really bothering me, and I was praying to God that the situation should not culminate into a physical capture in this instance; otherwise, there would be lots and lots of civilian casualties. Since the special court of the Central Bureau of Investigation had passed the order for the administration to take over the Dera, it had to be vacated. While I was thinking about the task at hand, the primary query that struck me was the degree of urgency of implementation of the court order, and it was then that there was a spark in my eyes. In the Army, a task is generally qualified by the time stipulated for its completion, but in this case, there was no time specified to the best of my knowledge. However, to be certain, I thought of calling Rajeev Ranjan, Commissioner, Hisar Division, since Sirsa district came directly under his jurisdiction, and I knew him personally. Haryana state was divided into six administrative divisions and Hisar division directly controlled the zones of Fatehabad, Jind, Hisar and Sirsa. Rajeev Ranjan was delighted to get my call and complimented me for restoring the law-and-order situation in Sirsa. He was particularly appreciative of the fact that there was not a single casualty since the time the Army had taken control of the situation. This was an achievement since prior to our arrival, there had been six casualties as a result of firing by the paramilitary force on a mob that had attacked them. After the initial pleasantries, I shared the purpose of my call—to seek clarity on the urgency for the clearing of the Dera in Sirsa. I was happy to learn that while the court had not specified any time, the court order had to be implemented at the earliest. I could sense a window of opportunity since, in the Army, the earliest possible time in normal circumstances is best left to the man on the ground who is executing the task.

Now I started to analyse the options available under the prevailing circumstances. After a lot of deliberation, I could home in on two options. My mind was advocating the first option—to go ahead with the momentum of the operation without wasting any time. In actual operations, success is directly proportional to the momentum of the operation, and the tempo of the operation must never die down till the task is completed. Accordingly, it translated into the execution of phase two, which is the clearance of the Dera immediately on completion of phase one, which is the cordon around the Dera. This was going to cause several casualties to the people inside the Dera and possible collateral damage to the Dera itself, which would hurt the sentiments of its followers. However, my heart was stuck with the second option to try and peacefully vacate the Dera by giving a chance to the followers to come out on their own. The second option required an operational pause after the completion of the cordon around the Dera in phase one. There was a bleak chance that the followers may decide to come out of the Dera on their own. The second option, though with fewer chances of success, was much more humane. I was not in a hurry to make the final choice since phase one was still underway, and I had time till the next morning to pass the executive order for phase two of the operation. In the morning, inside the Joint Operations Room, everyone was surprised to see a Military General talking like an administrator, wherein I gave clear instructions to plan for the second option. I had even asked for breakfast and the buses to be parked outside the Dera to facilitate travel for the people coming out of the Dera. After ensuring that every instruction of mine was followed, I walked into the Dera and addressed the people inside with absolute clear orders to vacate the Dera in the next one hour or the Army would do serious business. I won't lie, there came a moment in the process where I trembled thinking that if no one stepped out, I would have to take the most dreaded decision. Minute by minute, people around me started getting disillusioned with my plan, but my eyes were glued to the main iron gate with concrete pillars on either side. It was around noon with the sun over our head, and there, breaking the sunbeam, I discerned a white shade in the joint of the gate. With a screeching noise, the gates opened, and a man in his mid-fifties with a white turban on his head stepped out. Even though his face expressed anxiousness, for me he was the golden light, and the rest is history!

22

HASTEN SLOWLY

While I was posted at Along in Arunachal Pradesh, the locals would always talk about the classic example of how the Ganges River is a symbol of worship, while the Brahmaputra River is a symbol of terror. They would explain it in a manner that actually would sum up the philosophy of life. Both these rivers or their tributaries originate at Tibet, and as we all are aware, the Ganges flows into India and the Brahmaputra flows for the most part of its journey in China, where it is known as 'Tsang Po', and finally, it enters India near Tuting in Arunachal Pradesh. India welcomes the mountain river with a new name, 'Siang', and finally, it merges with the rivers Lohit, Siyom and Subansari to take the shape of the mighty fury of the Brahmaputra. Once again, both the rivers, Brahmaputra and the Ganges, meet each other in the Bengal Delta, which is popularly referred to as the Sunderban Delta, just prior to discharging into the Bay of Bengal. To sum it up, both the rivers are born together, take on the journey of life on different paths, and finally, the last leg of their journey brings them together again. Now let us analyse what these two rivers actually do in their journey of life! So many civilizations flourished on the banks of the Ganges, whereas the Brahmaputra kept changing its course every year and caused floods and devastation in areas adjacent to its course. Therefore, we worship the Ganges for her good deeds and Brahmaputra is labelled with only a six-letter word—'terror'. The same holds for life, too—it is not important where you are born, but it is more important to do good deeds in the journey of life. The classic example of these two rivers had a major impact on my life, and I would request each one of you to ponder over this divine act of God.

Something which we all have learnt in our school days but have conveniently forgotten thereafter is Newton's third law of motion, which talks about action and reaction being equal and opposite. I call it the 'law of life', something I have spoken about earlier in this book; whatever we do in our own life gets back to us in the same proportion. Therefore, remember that someone up there is balancing everything that we do, and it is in our own interest that we do good. Well, while talking about good deeds, I would also like to raise the issue of right and wrong. How do you decide if what you are doing is right or wrong? It has taken me almost my entire life to solve this puzzle, and I call it a puzzle since the same act at times is right and in another situation is wrong. Killing a human being is the ultimate crime, but when you kill him on the border, you are rewarded with medals. Even when you go back in history, there are innumerable examples which make you wonder about right and wrong. We all know how Dronacharya in the epic, *Mahabharata*, was killed by a false rumour that implied his son Ashwatthama was no more, but the man who announced Ashwatthama's demise was Yudhishthira himself. We also know how Dronacharya asked for Eklavya's thumb to ensure that Arjuna became the best archer. The more you read about these mythological stories, the more confused you will be regarding what is right and what is wrong. It has taken me the past three decades to arrive at a simple formula for quantifying right and wrong; whatever you do in life has to be judged as right by your own conscience. You are your own best judge; you and only you can decide whether what you are doing is right or wrong. Therefore, in life, stop worrying about what people say and simply take a conscious call. I can assure you that your conscience will always guide you on the right path. I remember a story narrated by my grandfather about a father and his son going on a voyage. In the first village they reached, the people ridiculed the son since he was sitting on the donkey and his old father was walking, and so the son disembarked and asked his father to ride the donkey. While in the next village, the people ridiculed the father for being unkind towards his son since he was seated on the donkey and poor son was walking. The father then decided to sit on the donkey along with his son. Still in the next village, the people ridiculed both father and son for being seated on the donkey and showing no mercy towards the poor donkey; and finally, everyone ridiculed the father and son for walking while the animal was being unutilized.

Therefore, no matter what you do in life, you will be criticized, and you have to accept it, but you must continue doing what you feel is best under the circumstances. Many a time, you may have to change your stance, but remember to never compromise on your core values, just like the 'round stones in the river bed'. Let me explain this. To survive the flow of water, a stone has to change its shape. However, it never compromises on its core values, and in fact, it becomes harder than usual when it comes to its core values. That is why they say that what survives in the river bed is a round stone, and all other stones in various shapes will be washed away due to the heavy water current. Similarly, in life (do not take me the wrong way if I may say that in today's world) first learn to survive; to bring in any effective change, you may have to compromise on your shape many a time, but never compromise on your basic core values. Be a part of life while being a spectator to see how events or situations unfold around you. Do not get too engrossed with life and do not always question the events unfolding around you. In case things happen your way, you are responsible for whatever is going to happen thereafter. However, in case events unfold in 'His way,'[65] he is responsible for whatever is going to happen thereafter. The essence of what I want to say is that events are not going to unfold the way you want them to, and therefore, you should not allow activities around you to disturb you. This is what I would refer to as 'unconditional happiness'.

There is no better way to live your life and there is no better literature in the world to guide you as to how to live life than the Bhagvad Gita. Please read it as many times as you can; you must own a personal copy. I personally do not associate the Gita with only one religion; rather, I consider it to be the greatest blessing on mankind, which teaches you how to live life. Start reading it if you have not read it already and start reading it again if you have already read it. Anything that happens around you will never disturb you once you read it regularly and constantly keep reminding yourself of its preachings. The Gita is the best way to lead your life and each word has hidden treasures in its meaning. I am refraining from trying to explain to you the teachings of the Gita, simply because I want you to discover this ocean of knowledge and wisdom on your own and in your own unique way. The first reading of the Gita may enlighten you

---

65   The Almighty.

with knowledge, but continuous readings of it shall impart you with *wisdom*. It is important to be wise in life rather than be knowledgeable. When you look back at your life, you would find many incidents where you must have done something which you never wanted to do. How do you justify that? If you were never in favour of doing what you did, then why did you do what you did? I can say that even your karma[66] is destined. Does this mean that you just sit down and be a spectator to whatever is happening around you? You cannot do that even if you want to do it. You will continue with your karma in life, but do not get perturbed with the end result of your action. It may be so because the Almighty wanted you to do what you did. Therefore, carry on with your life and be happy with whatever is happening around you, believe in the simple fact that whatever is happening is destined and will happen, irrespective of you wanting it or not wanting it to happen. In the same breath, I am also saying that do good deeds in life, but do not do anything which is against your conscience.

I wish to touch upon one more aspect of life once again before I close. It is very simple to say but equally challenging to follow in reality. 'Keep it simple' when it comes to life. Advancements in technology may have simplified the way to the moon but it has certainly not made our life any less complicated. You are still struggling to remember the number of passwords which are required

---

66 Action.

for your daily functioning, and in fact, some of you have designed another password to store your passwords. I am not against technology, but what I believe is when it comes to life, keep it simple. As I stated earlier in this book, life has to be a mixed bag and it definitely cannot be a straight line. Therefore, learn to take it as it unfolds and never be disturbed with the 'folds of life,' since without these folds, life would be dry and straight, and maybe at the threshold of being boring and dull. The best technique to iron these folds is by ignoring them, and trust me, the day you learn this art, life would be so very interesting, and you would be the happiest person on earth! Do not be like the deer who dies as a result of exhaustion because it is continuously searching for the fragrance of 'kasturi',[67] finally only to realize that the fragrance was within his own body! Similarly, happiness is within you, and the earlier you realize this fact, the better it is. Everybody says that 'happiness is a state of mind,' but I am of the firm opinion that you cannot be happy till you own the key to your 'state of mind'. Unfortunately, each one of us looks for happiness outside us, or plans to be happy tomorrow, which never comes. Therefore, it is my sincere request to each one of you to own this precious key to your happiness at this very moment so that you can be happy today, tomorrow and forever! Towards this singular objective, may you hasten slowly in life! May God bless you all.

---

67   Musk.

ACKNOWLEDGEMENTS

I would fail in my duty if I do not thank Shri Gugan Ram, who was my physical training instructor at Sainik School, Chittorgarh. He had served in the Indian Army, and post his retirement, he was training children to join the prestigious National Defence Academy, Khadakwasla. He was my inspiration and always motivated me work hard towards my dream of joining the Army. Hailing from the same region as me, he would not miss any opportunity to motivate and guide me in my training to qualify the toughest written examination of the Indian Army.

Words fail me to express my gratitude towards Injo Gakhal, my divisional officer at the National Defence

Acknowledgements

Academy, without whose care and support I probably would have been thrown out of the Academy on disciplinary grounds. Being the tallest cadet in the Academy, I was always under the watchful surveillance of my instructors, and Captain Gakhal would often shelter me like a godfather. Moreover, every punishment I received would simply disappear from the Battalion Routine Order. Captain Gakhal ensured that I passed out with my batch from the Academy; and as a parting gift, on the last day, he hugged me with a lot of warmth and affection—a gift I cherish to date!

My wife, Anita, thank you for bearing with me and being my anchor through this military journey. You accepted being second to the olive green while keeping me as your numero uno. I promise to always walk this journey of life with you by my side.

I am grateful to Damini for being the pillar of support who helped to bringing this book to life. It is only because of her conviction that this book is getting published by the best publisher in the world—HarperCollins. Your faith in me is inspiring!

Finally, I am grateful to Swati Chopra, my editor, for trusting my story and making it a part of the HarperCollins family. Swati is the sole reason that this book would reach millions of homes all over the world. The reputation of a thorough professional attached to Swati's name and her first compliment, that the story is very powerful, will always bring a smile on my face!

ABOUT THE AUTHOR

Major General Rajpal Punia, YSM, an alumnus of Sainik School, Chittorgarh, was selected for the National Defence Academy, Khadakwasla at the age of sixteen. He was commissioned into the Indian Army on 9 June 1984 and joined 14 Mechanized Infantry (16 Jammu and Kashmir Rifles). General Punia was the orchestrator of Operation Khukri while commanding a company as part of a United Nations Peacekeeping Mission in Sierra Leone. He commanded the oldest brigade of the Indian Army along the Line of Actual Control against the Chinese. He attended the National Defence College course in Thailand,

About the Author

and the United Nations Senior Mission Leaders course in Japan. He has the honour of commanding the prestigious Armoured Divison at Hisar. Under his command, he successfully controlled the Jat Agitation of 2017 and also vacated the Dera of Ram Rahim in Sirsa. He has had an illustrious career in the Indian Army spanning thirty-eight years. He is a fine orator and has been penning down his military experiences since the very beginning of his career. He is the author of the book, *Operation Khukri: The True Story behind the Indian Army's Most Successful Mission as Part of the United Nations*, published by Penguin Random House India in 2021, which is soon to be made into a motion picture. He has also authored the book, *A Soldier Speaks: Dil Se*, published by Notion Press in 2022. The General Officer has numerous articles and research papers to his name.

30 Years *of*
HarperCollins *Publishers* India

At HarperCollins, we believe in telling the best stories and finding the widest possible readership for our books in every format possible. We started publishing 30 years ago; a great deal has changed since then, but what has remained constant is the passion with which our authors write their books, the love with which readers receive them, and the sheer joy and excitement that we as publishers feel in being a part of the publishing process.

Over the years, we've had the pleasure of publishing some of the finest writing from the subcontinent and around the world, and some of the biggest bestsellers in India's publishing history. Our books and authors have won a phenomenal range of awards, and we ourselves have been named Publisher of the Year the greatest number of times. But nothing has meant more to us than the fact that millions of people have read the books we published, and somewhere, a book of ours might have made a difference.

As we step into our fourth decade, we go back to that one word – a word which has been a driving force for us all these years.

Read.